Evolutionary Studies

Everything is history, starting from the Big Bang or the origin of the universe to the present time. This historical nature of the universe is clear if we look at evolution of organisms. Evolution is one of most basic features of life which appeared on Earth more than 3.7 billion years ago. Considering the importance of evolution in biology, we are inaugurating this series. Any aspect of evolutionary studies on any kind of organism is a potential target of the series. Life started at the molecular level, thus molecular evolution is one important area in the series, but non-molecular studies are also within its scope, especially those studies on evolution of multicellular organisms. Evolutionary phenomena covered by the series include the origin of life, fossils in general, Earth–life interaction, evolution of prokaryotes and eukaryotes, viral and protist evolution, the emergence of multicellular organisms, phenotypic and genomic diversity of certain organism groups, and more. Theoretical studies on evolution are also covered within the spectrum of this new series.

More information about this series at http://www.springer.com/series/15220

Motoo Kimura

My Thoughts on Biological Evolution

Springer

Motoo Kimura (1924–1994)
National Institute of Genetics
Mishima, Japan

Translated by
Yoshio Tateno
National Institute of Genetics
Mishima, Japan

Kenichi Aoki
The University of Tokyo
Tokyo, Japan

ISSN 2509-484X ISSN 2509-4858 (electronic)
Evolutionary Studies
ISBN 978-981-15-6164-1 ISBN 978-981-15-6165-8 (eBook)
https://doi.org/10.1007/978-981-15-6165-8

Translation from the Japanese language edition: Seibutsushinka wo kangaeru by Motoo Kimura © Akio Kimura, published by Iwanami Shoten, Publishers in 1988. All Rights Reserved.

This Springer imprint is published by the registered company Springer Nature Singapore Pte Ltd.
The registered company address is: 152 Beach Road, #21-01/04 Gateway East, Singapore 189721, Singapore

In Memory

Motoo Kimura wrote this book in Japanese more than 30 years ago, to serve as an in-depth and then up-to-date introduction to evolutionary biology for students and young researchers not necessarily specializing in this field. Chapters 1–5 deal with the history of evolutionary studies, including concise but thorough accounts of paleontology, systematics, mutation, and natural selection. Chapters 6 and 7 delve more deeply into population genetics and molecular evolution, which are areas of research to which Kimura himself has made significant contributions. Chapter 8 focusses on the neutral theory of molecular evolution, and Chap. 9 is a speculative look at the future of mankind. These last two chapters strike me as being perhaps a little too optimistic, in that everything in evolution is explained within the framework presented in this book.

I worked with Kimura for about half a century. William B. Provine, a noted historian of science, once remarked to me that he had never known two scientists with quite different opinions on their research problem to maintain such a long collaboration. Kimura liked simple and elegant theory, and I remember that he admired theoretical physicists such as Richard Feynman and Freeman Dyson. On the other hand, I was a typical biologist. As a result, we sometimes disagreed on the interpretation of the data on molecular evolution and polymorphism. In particular, he was strongly attached to his simple neutral theory, which posits that mutations are either neutral or definitely selected. I thought that molecular evolutionary processes could not be so simple, i.e., natural selection acting on mutations at the molecular level could not be as simple as all or nothing. So the two of us often had heated discussions on this topic, with James F. Crow once in a while joining in the fray.

In the Analects of Confucius, we find words that loosely translate as: "Maintain harmonious relations with colleagues, but do not forfeit intellectual autonomy." Kimura and I worked together for a long time while disagreeing on the interpretation of data and engaging in heated discussions. Mutual trust and the frank and open-minded discussions based on this mutual trust were what made this long collaboration possible. Kimura, who held the senior position in the Department of Population Genetics, treated me as an independent researcher, whereas in some laboratories in Japan, professors are overly powerful and do not treat younger faculty as such. I am grateful to him for this.

There has in recent years been remarkably rapid progress in almost every field of bioscience at the molecular level. Particularly noteworthy is the clarification of the processes involved in the gene regulation of various tissues and organs, including epigenetic mechanisms; highly complex molecular machineries are connected directly or indirectly, yielding a well-organized system as a whole. To reiterate, this book was written more than 30 years ago, before these and other advances were made. Nevertheless, it remains relevant even today, because Kimura's work laid the foundations for what followed in molecular evolutionary studies. I am happy to see it translated into English—the translators, Yoshio Tateno and Kenichi Aoki, were junior faculty members at the time Kimura wrote this book—so that more readers may benefit from the message it carries.

National Institute of Genetics
Mishima, Japan

Tomoko Ohta

Translators' Notes and Acknowledgments

This translation was done on biweekly visits to the National Institute of Genetics, where we had access to the late Professor Motoo Kimura's private library. The translation is complete and faithful to the Japanese original except for the following: (1) the sex of a researcher has been suppressed as being irrelevant; (2) population labels have been revised in Fig. 8.1 to conform to current usage (the original terms have been retained in the text); (3) we have deleted the Japanese language literature (many of which are unavailable) from the reference list (they, however, are mentioned in the text); (4) Muto et al. (1985) in the References has been replaced by Yamao et al. (1985) to be consistent with the citation in the text.

We thank Professor Naruya Saitou and his secretary, Mrs. Masako Mizuguchi, for their patience and hospitality during the 2-year period. We are also grateful to Naruya Saitou for acting as our "agent" in negotiations with (1) Mr. Akio Kimura, who holds the copyright, and with (2) the two publishers involved, Iwanami which published the Japanese language original and Springer which has now published this English language translation. We thank Akio Kimura for his permission, Iwanami for providing the original photographs which have been reproduced here, and Masako Mizuguchi for tracing the original figures.

Mishima, Japan
Tokyo, Japan

Yoshio Tateno
Kenichi Aoki

Preface

Research on the evolution of organisms may be one of the perennial tasks in biology. Questions in evolution are related to almost all areas of biology; whenever a new field was pioneered in biology, the new findings obtained there have led to a deeper understanding of evolution. It may be quite rare that a researcher in biology has not taken an interest in evolutionary questions at least once during his/her entire career.

In retrospect, when I was a college student majoring in botany some 40 years ago, some people openly stated that "serious biologists do not study evolution"; evolutionary research did not have a high standing in scientific circles. Generally speaking, discussions on evolution often were speculative and futile, which is sometimes true even now; but with developments in population genetics and molecular evolution, it is true that the situation has improved. In fact, research on evolution has reached the stage where it deserves to be called "evolutionary studies."

I have devoted myself for a long time to the theoretical (mathematical) study of population genetics, have had a profound interest in the mechanisms that drive evolution, and have familiarized myself with the literature on evolution based on genetics. From the 1950s to the early 1960s, I was one of the adherents of the "synthetic theory of evolution," during the period when it was fashionable as the established theory. However, as I incorporated knowledge on molecular genetics into classical population genetics, set myself the goal of constructing a new of theory of population genetics, and eventually 20 years ago proposed the neutral theory of molecular evolution, I changed my position from panselectionism to one that recognizes the importance of chance effects.

In this book, I have summarized the current state of evolutionary studies, with the general reader in mind, and have tried to include most of the important issues that are currently under discussion in this area. As a result, this book has unavoidably grown too long for a pocket edition. Nevertheless, I have made an effort to make it readily comprehensible to students in the humanities with an interest in biological evolution and have tried to weave in some original thoughts to make it a distinctive book. For example, in Sect. 8.5 of Chap. 8, I have discussed one of the major problems in today's evolutionary studies, i.e., "how to relate molecular evolution to phenotypic evolution." For this reason, this book has been written to contain material that should be of interest even to experts in evolutionary theory, and I hope that it will be read by a wide range of readers. Moreover, I believe that this short book may be useful as a textbook depending on how it is used.

At present, evolutionary theory is experiencing a boom, and many books are being published on the subject. However, evolutionary theory has always had aspects of a quagmire, and I pray that this book may serve as a guide for young readers to avoid becoming stuck. In addition, I would be happy if I can convey through the various topics treated in this book, how a new academic discipline is created.

I remember that the late Dr. Taku Komai, my mentor to whom I am greatly indebted, published late in his life a tome entitled "*Evolution of Organisms on the Basis of Genetics* (Baifukan, 1963, in Japanese)." This is a great book, which is even now sufficiently informative and which was published when Professor Komai was 77 years old. In the preface to this book he wrote, "I would like to dedicate this book as a token of my gratitude to my mentors, colleagues, friends, and family for their guidance, encouragement, support, and assistance throughout my life." I am retiring this spring from the National Institute of Genetics where I have worked for many years, and although I am 14 years younger than Professor Komai was at that time, it is with the same sentiments that I dedicate this book to Professor Komai and the many people to whom I am indebted.

The occasion for writing this book was provided by Mr. Shigeki Kobayashi of Iwanami Publisher when some 10 years ago he asked me to write a book on biological evolution. Subsequently, I promised to put together a pocketbook, but was too busy to bring it to fruition. I am very happy to have completed this task at last. Mr. Nobuaki Miyabe of Iwanami Publisher, the editor in charge of this book, carefully read each chapter of the manuscript and made various suggestions that helped to make this book comprehensible to the general reader. In addition, in regard to Chap. 9 where I discuss the future of mankind, I am grateful to Mr. Yoshimasa Yoshinaga, a science journalist, for useful advice on how to convey my intentions clearly. Lastly, Mrs. Yuriko Ishii, who prepared the manuscript of this book, has worked for me for many years with devotion and self-sacrifice. I take this opportunity to thank her with all my heart.

Mishima, Japan
February 1988

Motoo Kimura

Contents

1 Diversity of Organisms and Views on Evolution 1
1.1 Diversity of Organisms 1
1.2 Biological Evolution as a Fact 2
1.3 History of the Development of Evolutionary Theory 4
1.3.1 Lamarck and Darwin 4
1.3.2 Contribution of Mendel 10

2 History of the Development of the Theory of Evolutionary Mechanism on the Basis of Genetics 15
2.1 Troubled Beginnings 15
2.2 Formation of Population Genetics 18
2.3 Synthetic Theory of Evolution and Panselectionism 27
2.4 Studies of Molecular Evolution and the Neutral Theory 30
2.5 Other Evolutionary Theories 33

3 Tracing the Course of Evolution 35
3.1 Outline of the History of Life 35
3.2 Evolution of Vertebrates 39
3.3 Evolution of Mammals 43
3.4 Evolution of Primates and the Emergence of Hominins 45

4 Mutation as an Evolutionary Factor 49
4.1 A Genetic View of Life 49
4.2 Nature and Variety of Mutations 52
4.3 The Nature of Gene Mutation 57
4.4 Phenotypic Effects of Gene Mutations 60

5 On Natural Selection and Adaptation 65
5.1 Darwin on Natural Selection 65
5.2 Modern Developments in the Theory of Natural Selection 72

6 Introduction to Population Genetics . . . 85
6.1 What Is Population Genetics? . . . 85
6.2 Gene Frequency and Mating System . . . 87
6.3 On Genetic Equilibrium . . . 91
6.4 On Genetic Drift . . . 97
6.5 Behavior of a Mutant Gene in the Population . . . 100

7 Introduction to Molecular Evolution . . . 103
7.1 Eve of Molecular Evolutionary Studies . . . 103
7.2 Basic Knowledge for Understanding Molecular Evolution . . . 104
7.3 Estimation of the Rate of Molecular Evolution . . . 107
7.4 Characteristics of Molecular Evolution . . . 110
7.5 Accumulation Process of Mutations within a Species . . . 114

8 The Neutral Theory and Molecular Evolution . . . 119
8.1 Explanation by the Neutral Theory . . . 119
8.2 Intraspecific Variation at the Molecular Level . . . 124
8.3 Molecular Evolutionary Clock and Molecular Phylogeny . . . 127
8.4 Other Topics Related to Neutral Evolution . . . 130
8.5 Bridging Molecular Evolution and Phenotypic Evolution . . . 134

9 An Evolutionary Genetic World View . . . 139
9.1 The Human as a Product of Evolution . . . 139
9.2 Thinking about the Question of Eugenics . . . 142
9.3 Positive Eugenics and the Future of Mankind . . . 145
9.4 Human Expansion into Space and Evolution . . . 147

References . . . 151

Diversity of Organisms and Views on Evolution

1

1.1 Diversity of Organisms

The Number of Species on the Earth There is truly a great diversity of organisms living on the Earth. It is said that the number of species recorded so far may be as many as one million and several hundred thousand. Among these, more than one million are animals, of which the majority (approximately 700 thousand species) are insects. The kinds of plants including fungi and algae number about 500 thousand. In addition, more than 2000 species of bacteria (including cyanobacteria) are known.

Of course, there are likely to be a good number that remain undiscovered, and it is an interesting question how many species exist on the Earth in all. While the animals and plants that have been recorded so far are mainly from temperate zones, there is actually a greater abundance of biological species in the tropics. In fact, it is estimated that two to three times more species live in the tropics than in the temperate zones. Thus, it is reasonable to expect that 3–5 million species inhabit the Earth if we include undiscovered ones. However, based partially on surveys of insects living among the leaves in the treetops of the tropics, it has recently been proposed that more than 10 million species may exist on the Earth. On the other hand, the tropical rain forests which are treasure houses of biological species are currently being destroyed by development with excessive vigor, and it is said that species are being lost at the rate of about 1% per year. Thus, not a few scholars warn that unless we act quickly, almost all tropical rain forests will disappear from the Earth a hundred years from now. This loss will be accompanied by the extinction of many species before they are recorded.

Adaptation and Variation Organisms inhabit diverse environments; on the one hand, there are algae that actively proliferate on the snow of high mountains at temperatures near 0 °C; and as an extreme example, there are bacteria and fungi that live in salty ponds at −23° in Antarctica. Moreover, and on the other hand, bacteria (heat resistant bacteria) are known that live in hot springs at 90°, which are too hot to dip one's hand in. Organisms survive on high mountain tops as well as at the bottom

M. Kimura, *My Thoughts on Biological Evolution*, Evolutionary Studies,
https://doi.org/10.1007/978-981-15-6165-8_1

of deep seas. Just comparing body size, while the blue whale measures 30 m in length, there are bacteria as small as one micron (one thousandth of one millimeter), and there are bacteriophages infecting and causing diseases in bacteria that are even smaller.

It is well known that these various organisms lead their lives superbly adapted to their environments. The ingenuity of the adaptive mechanisms, from behavior to the specificity of chemical substances, amaze us the more they are examined. However, individuals of the same species, although they are well adapted, are not entirely identical, and in most cases there are differences, i.e., variation among the individuals. In fact, variability and adaptability are especially remarkable attributes of organisms.

Achievements of Molecular Biology The study of the diversity of organisms originally formed the basis of classical studies in biology including taxonomy, but due to the striking developments in molecular biology of recent years, its status has undergone a large decline. Recently however, research on molecular evolution is being actively pursued, and I believe that the study of the diversity of organisms, i.e., systematic biology in the broad sense, shows indications of being rehabilitated. Moreover, it is an area that is indispensable for the development of biological resources.

Needless to say, the achievements of molecular biology (molecular genetics in particular) in clarifying the molecular mechanisms of replication, which is a property of organisms, and the material basis of life, and in deepening our understanding of the essence of life are extraordinary. As a result, it is now clear that the blueprints (commands) for building the bodies of higher organisms including us and maintaining life are written with four kinds of bases of DNA (A, T, C, G) in the nuclei of fertilized eggs, and furthermore, that the genetic code used there is essentially identical for all organisms on the Earth.

That molecular genetics should have in this way elucidated the unifying principles underlying biological diversity, and that it should have given us the possibility of manipulating genetic material and of modifying the genes of organisms will be counted by future science historians as one of the greatest scientific achievements of the twentieth century.

1.2 Biological Evolution as a Fact

Evolutionary Theory and "Creation Science" It was the English biologist Charles Darwin who first convinced scientists of the world of the fact that the various and diverse organisms on the Earth were not separately created by God, but achieved their present forms by gradual changes over a long time span from a few common ancestral organisms. The first edition of his famous book *The Origin of Species* was published in 1859, about 130 years ago.

That organisms evolve is now widely accepted as an indisputable fact in the same way that the Earth is round, and very few people in Japan doubt this. But, in the

United States, there are people of a Christian sect who espouse a doctrine that takes the Biblical Genesis literally. They strongly oppose evolutionary theory, and assert that the story of biological origins based on the Bible should be taught in public schools under the name of "Creation science" with equal weight as evolutionary theory, and they forcefully promote a campaign for this purpose. In some states this campaign is a formidable political force, and courtroom litigations are currently in progress, as is sometimes reported in newspapers and magazines. It is fortunate that in Japan, where the Christian tradition is not as strong as in Europe or the United States, there is no worry that society will be poisoned by such misguided ideas, at least with regard to biological evolution.

Evidence for Evolution Here, I would like to briefly touch on the evidence for biological evolution.

Firstly, the most direct evidence for evolution is obtained from research on fossils which are the remains of past organisms. By arranging fossils contained in different geological strata according to the ages of these strata, we can clearly discern the routes by which organisms gradually changed from the distant past to the present. It is also important that intermediate forms between various extant species are often found. For example, it is interesting that an early amphibian, *Ichtyostega*, had four short limbs and a fish-like tail, which represents a transitional form exactly like a frog with a fish's tail.

The second kind of evidence is obtained from research on the classification of organisms; when we group together similar organisms and compare them, it is found that the various organisms are not totally independent of each other, but are related to various degrees.

Thirdly, when the geographic distributions are studied, there are many aspects that cannot be explained unless we assume that species differentiated by gradually adapting to their environments.

Fourthly, when we compare organisms by morphology, physiology, development and behavior, we see many facts that can be fully understood only by recognizing that various species of organisms arose by evolution. For example, the fact that the human has gills during early development and are born as babies after a subsequent phase with a tail, can be readily understood if we admit that human ancestors evolved from fish with gills living in water and from animals like monkeys with a tail.

Fifthly, we have the evidence obtained from the comparison of informational macromolecules such as proteins and DNA. This will be discussed later in detail in Chap. 7 which deals with molecular evolution; as one example, the amino acid sequences of the blood pigment hemoglobin, when compared among various vertebrates, are seen to differ in proportion to their taxonomical relationships. This can only be understood if there was a common ancestral hemoglobin molecule, and these species of organisms diverged from each other by the intraspecific accumulation of mutations. The comparative study of DNA base sequences makes this even clearer. Moreover, the universality of the genetic code shows that all organisms on the Earth are derived from a common ancestral molecule.

1.3 History of the Development of Evolutionary Theory

1.3.1 Lamarck and Darwin

It is said that the germ of the idea that biological species are not invariant but rather change gradually over long time spans can already be found in the writings of the Greek philosophers before Christ, in particular Empedocles and Aristotle. However, these writings are only of interest from the perspective of the history of science, and have almost no relevance to evolutionary theory as a modern science.

Lamarckism and the Inheritance of Acquired Characters The first scholar to propose a scientific theory of biological evolution based on fact may have been Jean Baptiste Lamarck in France. In the early nineteenth century when almost all biologists believed that species were invariant, he proposed that organisms evolved from lower to higher. *Philosophie Zoologique* published in 1809 is well known. He holds an important place second to Darwin in the history of evolutionary science, in that he clearly recognized evolution as a fact. He is also the first scholar to submit a phylogenetic theory of evolutionary mechanism. He hypothesized that changes acquired by an individual during its life time were transmitted to the next generation as changes in inherited properties. For example with regard to the evolution of the giraffe's neck, his explanation was that the present day giraffe with a long neck and long limbs evolved as a result of the cumulative efforts of the ancestors of this animal continually trying to eat leaves on the high branches of trees, and this habit being repeatedly maintained (Fig. 1.1).

Fig. 1.1 Lamarck

Lamarck's evolutionary theory is sometimes called "the theory of use and disuse", in other words, it is based on "the inheritance of acquired characters". However, with the subsequent development of genetics, the inheritance of acquired characters has been completely discredited. Therefore, the so-called Lamarck's theory is not correct as a theory of evolutionary mechanism. Nevertheless, scholars who propose a Lamarckian theory of evolution are present even now, suggesting that the idea of the inheritance of acquired characters can be eliminated from our common sense only with great difficulty. The famous American paleontologist George Gaylord Simpson, in remarking on Lamarck's theory, has gone as far as to say that it is a pity such an attractive theory has been proved wrong (Fig. 1.2).

Lamarck is now regarded as a great biologist who first clearly recognized that organisms evolve, but it appears that he was not taken very seriously during his lifetime. According to Simpson, Lamarck was barely noticed when he published *Philosophie Zoologique* (he was already 65 years old at that time), lost his eyesight in old age, and passed away in miserable circumstances.

It appears that a severe attack by Georges Cuvier, who then dominated biological circles in France, on Lamarck's evolutionary thinking had a particularly adverse effect on his standing. For details on the life and writings of Lamarck, I recommend *History of Evolutionary Theory* (Iwanami pocket book, in Japanese) by Ryuichi

Fig. 1.2 Statue of Lamarck in the Paris Botanical Garden. Standing beneath are Professors Jacques Ruffié of the College de France (right) and Tomoko Ohta of the National Institute of Genetics (left)

Yasugi. This book by Mr. Yasugi is an excellent commentary on the history of evolutionary theory up to Darwin.

Neo-Lamarckism and Weismann's Experiment Lamarck became famous long after his death, in the aftermath of the great controversy that followed on the publication in 1859 of Darwin's *The Origin of Species*. Among the opponents of Darwin, there appeared a group that totally rejected natural selection and asserted that the direct effects of the environment were the prime movers of evolution. Their claim was drawn from Lamarck's book and hence labelled Neo-Lamarckism, but in reality it was different from Lamarck's theory. In any event, it is unfortunate for Lamarck that the word Lamarckism came to be used not in the meaning of a theory that organisms evolve, but rather to indicate the scientifically erroneous notion that acquired characters are inherited. Not many people now seem to realize that the word Lamarckism is used in a slightly different sense from Lamarck's evolutionary theory.

It was August Weismann who made a frontal attack on Lamarck's theory. He pointed out in a series of papers that it was not only unnecessary to assume the inheritance of acquired characters to explain evolution, but also that the assumption was not correct. It is well known that Weismann reported an experiment in which he repeatedly cut off the tails of newborn mice for 22 generations, without producing a reduction in the tail lengths of the mice that were born. There are not a few biologists who even now flatly deny that this experiment is meaningful in refuting Lamarck's theory, which goes to show that traditional evolutionary theory is divorced from experiment and observation, and has the property that it is free to interpret observed facts any way it chooses.

Development of Genetics and the Theory of the Inheritance of Acquired Characters This century [the twentieth] has seen major developments in Mendelian genetics with the addition, one after another, of new revolutionary insights, and culminating in the present state represented by molecular genetics; but not a shred of evidence in support of the inheritance of acquired characters has been obtained. It is surprising that there are still some scholars who nevertheless cling to Lamarck's theory and insist that evolution cannot be explained by Mendelian genetics. For example, one scholar claims that, when the environment (ambient temperature, etc.) changes over a long time span, due to its direct effect the genes themselves change through the inheritance of acquired characters over that long time span, and as a result biological evolution occurs. In this way, the argument is still seriously made based on the conviction that such things, although not verifiable as direct experimental fact, will surely occur over a long time span of several million years, which reveals the difficulty of evolutionary theory as a science.

The fact revealed by molecular genetics and which provides the basis for thinking about the inheritance of acquired characters is that genetic information is inscribed on DNA in the form of a base sequence; that this is transcribed into RNA; that the amino acid sequence of a protein is determined based on this; that a species-specific biological organism is formed; and that its life is maintained. The flow of information here is unidirectional from DNA to RNA to protein. Therefore it seems impossible that, due to the effects of environment and habit, the DNA base sequence in the nucleus of a reproductive cell should change, especially to better adapt the organism to the environment.

However, when reverse transcriptase which synthesizes DNA from an RNA template was discovered (1970), the claim was made that the inheritance of acquired characters could thus be explained at the molecular level. Especially famous is an experimental result announced by E. J. Steele and colleagues, that when immunological tolerance (the phenomenon by which immunological response to an antigen is suppressed) was induced in newborn mice, this was passed on to two descendent generations; this received worldwide attention. However, when other researchers later carefully repeated this experiment, they were unable to replicate the experiment of Steele and colleagues.

Difficulties of the Theory of the Inheritance of Acquired Characters What is more problematic is that to say that a change in a somatic cell due to the effects of the environment causes some genetic change, and that this is passed on to descendants, is not a substitute for Darwin's selection theory as an explanation of evolution. If the changes were all unsuitable for survival, they would be of no use.

In addition, one scholar has raised the following objection to the claim that acquired characters are inherited. In general, the morphological and physiological properties of an organism (in other words, phenotype) are not 100% determined by its set of genes (more precisely, genotype), but are also influenced by the environment. Moreover, the existence of phenotypic flexibility is important for an organism, and adaptation is achieved just by changing the phenotype. If by the inheritance of acquired characters such changes become changes of the genotype one after another, the phenotypic adaptability of an organism would be exhausted and cease to exist. If this were the case, true progressive evolution, it is asserted, could not be explained. This is a shrewd observation. Certainly, one of the characteristics of higher organisms is their ability to adapt to changes of the external environment (for example, the difference in summer and winter temperatures) during their lifetimes by changing the phenotype without having to change the genotype. For example, the body hair of rabbits and dogs are thicker in winter than in summer, and this plays an important role in adaptation to changing temperature.

In any case, when I read the writings of people who currently propound a theory of evolution that presupposes the inheritance of acquired characters, I receive the impression that they pay no attention to whether or not they are scientifically correct, but rather are sustained by "strong faith" and "wishful thinking".

Neo-Darwinism The evolutionary theory of A. Weismann was called Neo-Darwinism, and as its leader, he actively engaged in debate with the proponents of Neo-Lamarckism. Because of this there was a great deal of activity in evolutionary circles toward the end of the nineteenth century. Weismann was an extreme panselectionist, and he rejected all evolutionary factors proposed by Darwin except natural selection.

Darwin and *The Origin of Species* My narrative is slightly out of chronological order, but true scientific research on biological evolution, needless to say, began with Darwin. In 1859, half a century after the publication of Lamarck's *Philosophie Zoologique*, he published the famous *The Origin of Species* (he was 50 years old at that time). Using the enormous amount of data collected over many years during and after the circumnavigation of the globe on the Beagle, he not only convinced scholars of the world that biological evolution was a fact, but also showed that adaptive evolution occurs by natural selection. We can infer that this book had a large impact, not only in biology but also on human thought in general, from the fact that the physicist Ludwig Boltzmann once wrote that the nineteenth century would be remembered as the century of Darwin. Moreover, the famous geneticist Herman J. Muller, on the occasion of the centenary of the publication of *The Origin of Species*, called this book "the greatest book ever written by one person" (Fig. 1.3).

Fig. 1.3 Darwin (aged 72). On the veranda of his house about to take a walk along the "Sand Walk" (from The Autobiography of Charles Darwin, Collins, 1958)

As is well known, Darwin reasoned in analogy with artificial selection that selection would operate effectively in nature. This is because in all species many more offspring are born than can survive and, as a result, a struggle for existence ensues. Therefore, variation that confers even a small advantage in the survival of an individual is preserved, transmitted to the next generation by the "strong principle of inheritance", and gradually spreads through the species. Darwin emphasized the importance of the accumulation of beneficial variations of small effect, and concluded that new species arose and evolution occurred in this way (I will give a detailed discussion of natural selection in Chap. 5).

When Darwin published *The Origin of Species*, his greatest frustration was that the laws of inheritance by which genetic variation is produced and transmitted to descendants were unknown to him. Nevertheless, it was due to his genius and insight that he was able to understand the important role of natural selection in evolution. However, in the successive editions of *The Origin of Species* that followed on the first, he seems to have retreated from his position that natural selection is the main cause of evolution, perhaps because he gradually lost his confidence when faced with the severe criticisms against selection theory. At the same time, it appears his position changed to one that recognizes the important role played in evolution by the inheritance of acquired characters as well. Given the current situation where Darwin's theory of natural selection forms the mainstream of evolutionary studies and is regarded as sacrosanct by some, it may be difficult to imagine the severity of the storm of criticism directed at that time toward Darwin's theory.

A Paradigm Shift The remarkable aspect of Darwin's theory is that the idea of adaptation through natural selection, which forms its basis, has been further strengthened rather than weakened by the subsequent developments in genetics, and has become one of the most important foundations, in other words the norm, for interpreting biological evolution scientifically. The theory proposed by the American historian of science, Thomas Kuhn, that progress in the natural sciences proceeds by paradigm shifts, has recently become well known in Japan. From this standpoint the proposal of Darwin's theory clearly constituted a large paradigm shift.

Herein lies the big difference between Darwin's theory and scientifically useless arguments, such as that evolution is inherent in organisms, or that it is driven by some kind of vital force, or that organisms are fated to progress towards perfection. Recently, Dr. Kinji Imanishi in Japan has severely criticized Darwin's theory and orthodox evolutionary theory based on genetics, taking the firm stand that organisms "change because they are destined to change". It is surprising that quite a number of people follow his example and regard these words as deeply significant. However, it is meaningless to consider such babble as scientifically comparable to Darwin's theory. For example, it is as useless as to claim that cancer cells "arise because they are destined to arise".

1.3.2 Contribution of Mendel

The emergence and subsequent progress of genetics had the greatest influence on the development of evolutionary theory after Darwin. I will treat the development of the theory of evolutionary mechanism based on genetics in the next chapter, but before that I would like to tell you about Mendel, the founder of modern genetics.

Priest When Darwin wrote up his theory in *The Origin of Species*, the greatest difficulty he faced was, as mentioned before, that he did not understand the mechanism of inheritance. In fact, Darwin wrote in the first edition (1859) of this book, chapter 1 page 7, that "the laws governing inheritance are quite unknown". What is interesting here is that, while Darwin was writing this book, in the backyard of monastery in Brünn in Austria (currently Brno in Czechoslovakia) far from England where he lived, Mendel was diligently conducting breeding experiments with peas in pursuit of the laws of inheritance (Fig. 1.4).

Johann Gregor Mendel was born in 1822 in the small town of Hienzendorf in Austria. In 1843 at the age of 21, he obtained a job in a monastery in Brno, the capital of Moravia, and entered the priesthood. According to Professor Shingo Nakazawa who is well acquainted with Mendel's achievements, the breeding of agricultural

Fig. 1.4 Mendel

plants and domestic animals was then being actively pursued in the area centered on Brno, and the abbot of the monastery, F.C. Napp, who was a knowledgeable man, instructed Mendel to carry out research on the laws of inheritance. Mendel taught himself science, and at one time taught Greek and mathematics at a nearby gymnasium, and in 1850 took the qualifying examination to become a certified teacher, but failed. The abbot sent him to Vienna University for two years, which was useful for Mendel in acquiring much new knowledge. Later he taught natural science at the Brno Practical High School and again took the qualifying examination for a certified teacher, but it seems that he failed again.

Mendel's Laws Mendel began his experiments with peas in about 1854, continued the experiments for almost 10 years, and presented his results at the Brno Natural Science Society twice, in February and March of 1865. The content of this talk was published next year in a paper entitled *Experiments in Plant Hybridization*. What are today called Mendel's laws are to be found in this paper.

One of the characteristics of Mendel's research was that he focused, not on the plant as a whole, but rather on individual heritable traits such as stem height and seed form, choosing for each, such alternative traits as tall and short stems, or round and wrinkled seeds. Moreover, he crossed pure lines for these alternative traits. I will omit the details of the breeding experiments he conducted; but as one example, when the tall and short plants were crossed, the first filial generation were all tall; and when these plants were self-fertilized, in the next generation (second filial generation) tall and short plants were produced in the approximate ratio of 3 to 1.

Mendel explained this result as follows. Genetic traits are determined by genetic factors in the body (he used the term "element"); the factor for tall stems has an advantage (dominance) in the hybrids, and if we represent this by the upper case letter A, the short form is recessive and represented by the lower case letter a. These factors exist in the state Aa in the first filial generation, A suppresses a, and the plant becomes tall. Here, the factors A and a are particulate and are paired, do not mix, and when the gametes (pollen, ovules) are formed in the first filial generation, A and a segregate and only one of these enters each gamete at random. In other words, the probability is one half that a specific gamete receives A or that it receives a. Therefore, when a first generation hybrid is self-fertilized, the fraction of the next generation that receives A from both male and female gametes and becomes an individual AA is 1/4. Similarly, the fraction that receives a from both and becomes an aa individual is 1/4. Furthermore, the fraction that receives A from one source and a from the other and becomes Aa is 1/2. This result is often expressed as follows.

$$\left(\frac{1}{2}A+\frac{1}{2}a\right)^2=\frac{1}{4}AA+\frac{1}{2}Aa+\frac{1}{4}aa$$

Here, because A is dominant to a, the 1/4 that are aa are short, and the remaining 3/4 that are AA or Aa are both tall. Namely, tall and short plants are produced in the ratio of 3 to 1.

In genetics today, A and a are called "alleles", AA and aa are called "homozygotes", and Aa is called a "heterozygote". At the core of Mendel's laws is the claim that when hybrids produce gametes the factors A and a separate and enter the individual gametes, and this is now called "Mendel's First Law" or the "Law of Segregation". Also, due to the development of the chromosome theory of genetics, it was later found that genes reside on chromosomes and that alleles are located at corresponding positions of corresponding chromosomes (homologous chromosomes) derived from each parent. From this standpoint Mendel's Law of Segregation can be readily understood as the consequence of the homologous chromosomes moving to the two poles during meiosis which precedes gamete formation.

Incomplete Dominance In addition, due to the development of genetics after Mendel, it was found that a clear dominance relation does not necessarily exist between alleles, but that a condition called "incomplete dominance" can occur, where the heterozygote Aa is intermediate between both homozygotes AA and aa. In such a case it is preferable, in order to avoid misunderstanding, to denote the alleles not as A and a, but as A and A' or as A_1 and A_2. Also, the number of alleles is not limited to two, but it has been found that three or more, $A_1, A_2, A_3, \ldots$ may exist. It is well known that the human ABO blood group comprises three alleles I^{A}, I^{B}, and I^{O}.

Incidentally, in higher animals the individual has two sets of homologous chromosomes derived from the two parents and is diploid or in the diplo-phase state, and when a gamete (sperm, egg) is formed as a result of meiosis, one set of chromosomes enters it so that the gamete is in a haplo-phase state (haploid). And as most readers have learned in high-school, the diploid individual is reconstituted when sperm and egg are united by fertilization. However, in lower organisms such as bacteria the usual state of individuals is haplo-phase.

A Theory Ahead of Its Time Mendel's research was revolutionary and transcended the level of biology at that time. In particular, showing that the results could be predicted by computing probabilities on the assumption that the inheritance of various traits was controlled by factors was too far ahead of the times, and it is natural that ordinary biologists could not readily understand its importance. It took the efforts of scientists around the world over one century for the elements posited by Mendel to be conceptually established as genes occupying fixed positions on chromosomes, and for it to be ascertained that their nature was DNA. I would like to add one other thing, and it is that Mendel's explanation was the first example of the application to biology of the currently popular "probability model", and this may have been regarded by contemporary biologists as a much too simplistic argument, and not worthy of serious consideration. Come to think of it, it is wonderful that such a simple law is hidden behind the extremely complicated life phenomena.

It is said that Mendel's research was not publicly recognized for a long time, and received attention for the first time on being rediscovered by three scholars in 1900. An extreme opinion holds that Mendel's work was completely ignored and buried for 35 years, and suddenly gained acceptance when it was rediscovered. However, according to Professor Shingo Nakazawa, recent investigations have shown that the situation was apparently different, that Mendel's research was known across Europe, and that there were quite a number of people paying attention to Mendel's work. In fact, Mendel's research was described in the 9th edition of the world-famous *Encyclopedia Britannica* (1881). Without this, it is unlikely that the rediscoverers of Mendel's laws, C. Correns and others, would have known of the existence of Mendel's paper before publishing their research. The truth of the matter may be that Mendel's research was fairly well known internationally after its publication, but that no one appreciated its revolutionary importance.

Mendel's Tragedy Incidentally, in 1867, 2 years after the presentation of his research, Mendel was chosen to succeed Napp as abbot, and in the following year 1868 he was officially appointed. The Abbot of Brno Monastery at that time seems to have been a local celebrity and an important administrator. Mendel wished to conduct his experiments with plants even after becoming abbot, but this gradually became impossible because he was too busy. Later, in 1875 the Austrian government issued a law to impose a heavy tax on Catholic monasteries, and Mendel protested strongly against this. Mendel fought against the government officials with his characteristic stubbornness but was ultimately crushed. Eventually he became bed-ridden, and passed away on January 6th, 1884, a little before turning 62.

In my opinion, Mendel's tragedy was not, as is usually said, that he was not widely recognized while alive, but that he became an administrator for which he was unsuited, could not stick to a prudent policy nor engage in skillful political diplomacy, was defeated in his struggles against the government officials, and was never able to return to the research on plants that he cherished.

History of the Development of the Theory of Evolutionary Mechanism on the Basis of Genetics

2

2.1 Troubled Beginnings

Conflict Between Two Schools With the start of this century, Mendel's laws were rediscovered and modern genetics began its spectacular development. The problem annoying Darwin that the laws of inheritance were unknown was thus resolved. Logic would seem to dictate that Darwin's theory and Mendelian genetics be immediately integrated and that a theory of evolutionary mechanism based on genetics develop smoothly right away. However, the actual situation was quite different from expectation and full of vicissitudes.

In England, an intense almost emotional controversy immediately arose between the biometricians led by W. F. R. Weldon and Karl Pearson, who were academic descendants of Darwin, and the Mendelians represented by William Bateson. In fact, the conflict between Weldon and others, on the one hand, and Bateson appears to have preceded the rediscovery of Mendel's laws.

The biologist Weldon, who was influenced by Francis Galton—the latter was Darwin's cousin, the founder of eugenics, and also proficient in biostatistics—believed that statistical methods were optimal for conducting research on evolution. He tried to estimate the strength of natural selection by measuring various characters of plants and animals and their evolutionary rates. Pearson, who was Weldon's friend and a well-known applied mathematician, became interested in biological evolution under Weldon's influence and published many biostatistical (mathematical) papers on evolution and genetics. In retrospect, the laws of inheritance formulated by Pearson around that time were false, but the statistical methods (e.g., Chi-square method) that he developed through this research have made unexpectedly large contributions, not only to subsequent research in evolution and genetics, but also to research in biology and in science in general. Weldon and Pearson both followed Darwin in reasoning that evolution occurs gradually by the action of natural selection on small genetic variation.

By contrast, based on research on variation in plants and animals and in opposition to Darwin's claim, Bateson came to believe that it was not possible for evolution

M. Kimura, *My Thoughts on Biological Evolution*, Evolutionary Studies,
https://doi.org/10.1007/978-981-15-6165-8_2

to occur by the action of natural selection on continuous individual variation in a species. What was important and meaningful for Bateson was the manifestly discontinuous variation among individuals.

Victory of the Mendelians With the rediscovery of Mendel's laws, the conflict between the Mendelians and biometricians became even more acute. However, Weldon soon died young, experimental data supporting the truth of Mendel's laws were published one after another, and with the defeat of the biometricians becoming clear the dispute between the two schools ceased. In England, Bateson greatly contributed to convincing biologists in general of the truth of Mendel's laws. The term "genetics", which is now widely used throughout the world, was proposed by Bateson in 1906 to define the academic field of research on inheritance and variation (Fig. 2.1).

Here, I would like to mention one more thing, and it is that Weldon who attacked Mendel's laws nevertheless made an important contribution to research on natural selection. He measured snails and found that individuals with more or less than the average number of spirals had a lower viability than the average individual. It can be said that he is one of the first discoverers of the form of selection now called "stabilizing selection". Moreover, after Weldon's death, the Weldon prize was established at Oxford University in his memory, which is awarded once every 3 years to the scholar making the greatest contribution in the world to the field of biometrics. Recently Professor Tomoko Ohta of the National Institute of Genetics [Japan] was chosen as the awardee of 1986.

Fig. 2.1 The Weldon Prize medal with the portrait of W. F. R. Weldon engraved. Past recipients include R. A. Fisher (1929), J. B. S. Haldane (1938), and S. Wright (1947), the founders of the mathematical theory of population genetics

With the victory of the Mendel school, the number of biologists increased who were skeptical about the Darwinian thesis that evolution occurs by the force of natural selection acting on small continuous variation, and in its place the "mutation theory" of H. De Vries began to be widely accepted. De Vries is one of the rediscovers of Mendel's laws; according to his mutation theory a new species is not gradually formed by the action of natural selection, but arises at one bound by a sudden change in the genetic material. When this theory was proposed at the beginning of this century, it received the support of many biologists.

At present, it is thought that the "mutations" in the evening primrose discovered by De Vries are not mutations in the sense of the word as used today, but are likely to have arisen because this plant has complex chromosomal aberrations. However, De Vries' theory has done a great service in drawing the attention of many biologists to mutation as the true cause of genetic variation, and directing them to its investigation. Eventually, the existence of mutation was confirmed H. J. Muller, leading to the elucidation of its nature.

Resolution of the Contradiction As such, the first 10 years of this century saw active research aimed at resolving the question of whether natural selection acting on continuous small variation was as effective as Darwin had thought.

Especially famous is W. L. Johannsen's research on pure lines. Here, "pure line" refers to a genetically uniform strain of animals or plants. He used beans which propagate by self-fertilization (i.e., "selfing"), divided the beans that formed on the same plant or individuals of the same strain into two groups comprising heavy beans and light beans, and on examining the weights of the beans that formed on the plants raised from the beans of each group, found no difference on average. In general, he showed that selection had no effect within a pure line (1903). On the basis of this fact, he proposed the so called "pure-line theory", which attracted attention.

In retrospect, there is no mystery; as beans are self-fertilizing, the offspring arising from seeds forming on the same plant are genetically uniform; differences in weight among the seeds forming on a pure line individual are due to the environment (not due to a genetic difference); and it is natural that differences due to the effects of the environment are not transmitted to the next generation. However, with regard to the effectiveness of selection acting on continuous variation, Johannsen's research apparently played a role in intensifying uncertainty, rather than in resolving the problem. It was as it were a time of confusion. Incidentally, the term "Gen" meaning a gene was proposed by him.

But the confusion was gradually resolved, and it came to be accepted that there is no contradiction between Mendelian genetics and Darwin's theory. Contributing greatly to the resolution of this contradiction was the remarkable progress in Drosophila genetics made by T. H. Morgan and his school in the United States, as a result of which it became clear that genetic mutations exist with a very small phenotypic effect.

2.2 Formation of Population Genetics

Birth of Population Genetics Soon after, the efforts to place Darwin's evolutionary theory on a biometrical foundation based on Mendelian genetics bore fruit, and population genetics was born. A pioneering result is the so-called "Hardy-Weinberg law", which was published independently in 1908 by the English mathematician G. H. Hardy and the German doctor W. Weinberg, and which is well known even now (I will explain population genetics in some detail in a later chapter). Within 20 years of Hardy's and Weinberg's research, the population genetical implications of Mendelian genetics were elucidated by three scholars, R. A. Fisher and J. B. S. Haldane in England, and Sewall Wright in the United States, and by the early 1930s the mathematical theory of classical population genetics was mostly complete.

Let me say a few words about what population genetics is. Population genetics has as the target of its study the biological population, in particular a collection of conspecific individuals connected by sexual reproduction, or in other words a breeding society. In this population, various alleles exist in various proportions, which are called "gene frequencies". Population genetics investigates how these frequencies change under evolutionary factors such as mutation and natural selection. Needless to say, one of the important goals of population genetics is the elucidation of the mechanisms of evolution.

Fisher Among the three scholars, the one who had the greatest influence on the formation of the orthodox view of evolutionary genetics was R. A. Fisher. According to this orthodox position, the rate and direction of evolution are determined almost entirely by natural selection; mutation, migration, and random genetic drift have only an auxiliary effect. Here, random genetic drift is the phenomenon by which gene frequencies increase or decrease by chance over generations, and is more likely to occur the smaller the population is. I will explain this phenomenon later in a little more detail.

Many people refer to this orthodox position by the name of "Neo-Darwinism", which is the same as for Weismann's theory and therefore confusing, but essentially it derives from the fact that it is a position in accord with Weismann's tradition. As mentioned earlier, Weismann completely denied the inheritance of acquired characters and attributed the maximum importance to natural selection. In the United States, many scholars use the term the "Synthetic Theory of Evolution" in its place. The reason for this naming may be to emphasize their mentality, which incorporates all of the various evolutionary factors that are consistent with modern genetics in its treatment of evolution. However, in practice, their position was dominated by panselectionism (Figs. 2.2 and 2.3).

Fisher is an English mathematical statistician who is well known as the founder of modern statistics, but he was also responsible for epoch-making achievements in theoretical research on population genetics. In general, when a biological population reproduces, a virtually infinite number of male and female gametes are produced, a finite number are randomly sampled from among them, and the population of the

Fig. 2.2 Dr. H. J. Muller who pioneered the field of radiological genetics and also made significant contributions to the foundations of evolutionary genetics. Photo taken on the occasion of a lecture at the National Institute of Genetics in 1951

Fig. 2.3 Professor R. A. Fisher who is well-known as the founder of modern statistics and also made enduring contributions to the mathematical theory of population genetics. He visited the National Institute of Genetics in 1961 and lectured on human evolution. The author is on the left

next generation is formed from their union. He was the first to study this phenomenon, in which gene frequencies fluctuate by chance in this way, as a stochastic process. Here, a stochastic process is the mathematical formalization of probabilistic events that progress in time.

This is the already-mentioned phenomenon of random genetic drift, and Fisher used a heat-diffusion type partial differential equation to deal with it. He then obtained the result that in a population of N breeding individuals, the amount of genetic variation in the population (this is usually expressed as the variance of gene frequency, i.e., the square of the standard deviation) decreased by $1/(4N)$ in every generation. Here N is the quantity more precisely called the "effective population number" (this term was proposed by Wright); in general N is expected to be very large in various biological species, and he reasoned from this result that the rate at which genetic variation in a species decreases by random genetic drift is extremely small, and that on the other hand the action of natural selection is much stronger, so that the effect of drift was negligible in comparison. For example, if N is one million, the rate of decrease would be as small as 1 in 4,000,000 per generation. This calculation contained a small error, and it was later pointed out by Wright that the correct rate of decrease was twice this value, namely $1/(2N)$. Stimulated by this, Fisher studied this problem more deeply and produced a beautiful mathematical analysis. Either way, he maintained a completely negative position on the importance of random genetic drift in evolution.

In 1930 he published the immortal masterpiece, *The Genetical Theory of Natural Selection*, and made the greatest contribution to laying the foundations of natural selection theory on the basis of genetics. This book was for a long time like a bible in the field of the evolutionary mechanism theory. The strong tradition of excessive emphasis on natural selection in England likely owes much to this book by Fisher.

Haldane J. B. S. Haldane, along with A. I. Oparin, is known as the first proponent of a scientific theory of the origin of life; in addition, he was an outstanding scholar with a wide outlook who made many contributions in wide areas of biology. Although his contributions to the mathematical theory of population genetics were not as original as Fisher's, they had many notable aspects in the sense that they dealt with diverse subjects and were biologically relevant. In particular the series of papers by Haldane entitled *A mathematical theory of natural and artificial selection* beginning in 1924 were epoch-making at that time. In the introduction to the first paper, he states that a satisfactory theory dealing with natural selection must be quantitative. And he firmly believed that this was the only way to investigate the appropriateness or non-appropriateness of a genetical theory of natural selection (Fig. 2.4).

He investigated mathematically how various forms of selection influenced gene frequency change. For example, he calculated the effect of assuming alleles A and a, a viability of 1 for the dominant individuals AA and Aa, and a viability of $1 - k$ for the recessive aa individuals. Here, k is called the selection coefficient, and is often used in population genetics to designate the strength of selection acting on an allele. If k is 0.001 (in other words, if the dominant individuals have a 0.1% advantage in viability over the recessive individuals), he showed that for the allele A to increase in

Fig. 2.4 Professor J. B. S. Haldane made giant contributions, not only to the mathematical theory of population genetics, but also in a wide range of fields such as the origin of life, biological evolution, human genetics, and biostatistics. This photo was taken at Nikko when he visited Japan in 1956 to attend an international symposium on genetics

the population from 1% to 99% would take 16,483 generations. If k is 10 times larger, the number of generations required for the same change is one tenth of this.

Industrial Melanism in Moths Haldane applied this kind of theory to industrial melanism in moths that occurred in Manchester. Here, industrial melanism is the phenomenon in which, from about the middle of the nineteenth century, many kinds of moths in the industrial cities and their environs in England and countries on the European continent gradually changed from the original pale-colored form to the melanic form.

Of these, the most thoroughly studied is the peppered moth (*Biston betularia*); populations of this insect consisted almost entirely of pale-colored individuals until the middle of the nineteenth century, even in industrial cities. The patterns on the wings of the pale-colored form apparently serve as protective coloration against birds when these insects are resting on tree branches covered in lichen. At first the melanic form was merely a rare mutant form, but subsequently the melanic form increased in industrial cities, and had reached 98% by the survey of 1952–1956. In industrial areas the surroundings are blackened by soot, and it is thought that the melanic form is less likely to be discovered and eaten by small birds. However, even in species that have undergone industrial melanisation, the melanic form is still at a low frequency in rural areas not covered by soot and are very rarely seen in the mutant form. It can be said that the phenomenon of industrial melanism is the most striking example of evolution observed in a short period of time. Breeding experiments have shown that the melanic form is dominant to the pale-colored form and that the trait is controlled by single gene.

Haldane applied his calculations to these observations and estimated that the selection coefficient (k) had to be at least 0.33 and perhaps as large as 0.5 in order to account for such rapid change. Later, H. B. D. Kettlewell in England showed by detailed research in the laboratory and in the field that predation by birds would indeed exert selection of this strength.

Fixation Probability of a Mutant By the way, this kind of mathematical treatment of gene frequency change is deterministic and ignores chance fluctuations in gene frequencies. Nevertheless, there are not a few situations where this method is useful even now. Of course, his research was not limited to this, and he made important contributions to probabilistic treatments.

As an example, in 1927 Haldane succeeded under simple conditions in obtaining the probability that a mutant gene arising in a population would spread through the population under natural selection and be fixed (the frequency reaches 100%). In other words, when a single dominant mutant appears in the population and this has an advantage in viability of k over the preexisting gene, and assuming that the population is very large and random mating, he showed for the first time that the fixation probability of this mutant was approximately $2k$. Here, k takes a positive value much smaller than 1, and the population is sufficiently large. For example, if the dominant mutant allele (call this A) increases the viability of an individual carrying it by 1% relative to the preexisting allele (a), the fixation probability of this gene in the population would be 0.02.

Here, a word of caution is called for, which is that common sense suggests that a mutation advantageous for survival arising in the population would necessarily (with probability one) spread through the population, but that this is not so. The reason is that for several generations after its appearance, the probability is large that the mutant gene will be lost by chance from the population (the first individual is a heterozygote Aa, and the probability that A will enter a specific gamete produced by this individual is 1/2; since this individual will mate with another individual that it aa, it is easy to understand that with high likelihood there will be no individuals carrying A in the next generation.) Therefore if A has a 1% advantage over a, then even if one A appears as a mutation in the population, it will ultimately be lost with probability 0.98 and consequently not contribute to evolution (for fixation to be assured, it is necessary that A appear repeatedly). This fact is even now overlooked in many discussions of evolution. After Haldane, the problem of the fixation probability of a mutant has become important, and there has been much theoretical progress.

In 1932 Haldane published a book entitled *The Causes of Evolution*. In this book he summarized his research up to that time, discussed the mechanisms of evolution based on this and, together with the already mentioned book by Fisher published in 1930, contributed greatly to convincing biologists of the world that Mendelian genetics and Darwin's theory of natural selection did not contradict each another. Moreover, in a paper published in the 1930s, he noted that the mathematical treatment of evolution would become a respectable area of applied mathematics during the next half century; the truth of this prediction is now accepted by many biologists.

Wright In comparison to Fisher and Haldane, the distinguishing feature of S. Wright's research was his emphasis on the importance of genetic drift and non-additive interactions (epistasis) between genes. Epistasis means non-additive gene effects such that "one plus one equals two" does not hold, and in relation to Wright's evolutionary theory what is important is the following case.

Let us assume two pairs of alleles, *A* and *a*, and *B* and *b* (we assume that *A* and *B* are mutant genes at different positions on a chromosome or on separate chromosomes). In addition, we assume that *A* by itself has a small disadvantage to *a* for the survival of the individual, and similarly that *B* by itself has a small disadvantage to *b*. However, *A* and *B* together have an advantage over the combination of *a* and *b*; therefore the double mutant individual *AABB* is assumed to have an advantage in survival over the original *aabb* type individual. It is not easy to think of a concrete example of this, but let us suppose that the upper and lower jaws of an animal are controlled by independently acting polygenes, and that *A* makes the upper jaw larger than *a*, and that *B* makes the lower jaw larger than *b*. If this is the case, the upper and lower jaws must fit together well in order to chew food; *AB* represents large upper and lower jaws that fit together well, and we can imagine that it will have an advantage, certainly over *Ab* and *aB*, and over *ab* as well.

In his early work, Wright devised a calculation method called "path coefficients", which he used to produce epoch-making contributions on inbreeding and assortative mating, but what truly made him famous was the paper entitled *Evolution in Mendelian Populations* published in 1931. This paper is regarded as a document representative of classical population genetics, along with the afore-mentioned books by Fisher and Haldane.

Professor Wright and Myself This paper was a laborious work exceeding 60 pages in length, which appeared in the American genetics journal *Genetics*, but due to its mathematical content does not seem to have been read very much considering its fame. In 1948 about 3 years after the end of the war or a little earlier, I tackled this paper when I was a student in Kihara Laboratory at Kyoto University, and I still remember struggling to understand its import. I was impressed by the importance of this paper, and this decided the direction that my subsequent research would take. Later, I obtained a position at the National Institute of Genetics in Mishima, and then had an opportunity to study abroad in the United States, where for 2 years beginning in the early summer of 1954, I studied under Professor J. F. Crow in the Laboratory of Genetics at the University of Wisconsin. Around that time, Professor Wright retired from the University of Chicago and became professor of genetics at the University of Wisconsin, a few months after I had arrived. I feel lucky

Fig. 2.5 Dr. S. Wright from whom the author received the greatest influence in his work in population genetics (the white-haired gentleman second from left). The author is to the left of Dr. Wright, Professor J. F. Crow is to the right (facing backward). Taken in autumn of 1955 at the Genetics Laboratory picnic when the author was a graduate student at the University of Wisconsin

to have had the opportunity to attend his lectures and to enjoy his friendship (Fig. 2.5).

I was acquainted with Professors Fisher and Haldane (both deceased) but Professor Wright had an entirely different personality, mild and shy, and was an unpretentious person who one would not consider after a short chat to be a great scholar. He was born in 1889, which will make him 100 years old in about 2 years. Even now he is healthy both physically and mentally, and it seems certain that a symposium will be held in forthcoming 1989 to celebrate his 100th birthday. Not a few people predict that he will personally give an important scientific lecture at that time (On March 3, 1988, while this book was in proof, we sadly received the news that Professor Wright had passed away. It seems that on the previous Sunday morning, he had gone for a walk, slipped on an icy pedestrian walk, and broken his bones, resulting in this sudden death.) (Fig. 2.6).

I would like to add that recently (1986) his biography entitled *Sewall Wright and Evolutionary Biology*, written by W. Provine who is an American historian of science and highly knowledgeable in population genetics, was published by the University of Chicago Press. The distinguishing feature of this book is that it is not a rehash of previous works, but rather the product of much effort by the author Professor Provine, based on many long interviews with Professor Wright and on the examination of all copies of about 15,000 letters preserved by Professor Wright. It is my hope that this book will be read as a standard of research in the history of

Fig. 2.6 Taken 30 years later (in 1985) at Professor Crow's house. Dr. Wright is in the front row center, the author is on the left, and Professor Crow is on the right [Dr. Chung I. Wu is in the background]

science by many people, including in Japan, who are interested in evolutionary studies and its history.

Shifting Balance Theory Returning to the main subject, we can say that the task of rooting Darwin's theory of evolution in Mendelian genetics was mostly completed by the researches of the three great scholars Fisher, Haldane, and Wright during the 1930s. From 1932 onward, Wright proposed a unique theory of evolution, which he later called the "shifting balance theory". According to this theory, the most favorable condition for biological evolution is for a large population to be divided into many regional subpopulations. Under this condition evolution can proceed rapidly by the shifting balance process. Here the shifting balance process comprises the following three stages.

1. Genetic drift—within a regional subpopulation, random variation of gene frequencies occurs on a large scale due to small population size or fluctuation of selection intensity caused by environmental variation.
2. Individual selection within a population—within a regional subpopulation a combination of mutant genes that are individually disadvantageous but advantageous when coupled spread randomly, and when its frequency exceeds some threshold by chance, then selection at the individual level acts effectively, and individuals carrying this combination of genes spread rapidly to fixation within this subpopulation.

3. Selection between subpopulations—a subpopulation in which an advantageous combination of genes has been fixed by chance in this way will have a competitive edge over the other subpopulations, and will displace the surrounding subpopulations and spread gradually through the species.

When concentric expansion of subpopulations with such advantageous combinations of genes occurs at various localities, then where two such expansions meet a combination of genes emerges that is superior to either one, and this new combination will again spread concentrically from that point. In this way it is claimed that a virtually infinite number of advantageous non-additive gene effects (epistasis) is utilized in evolution. Incidentally, Wright's evolutionary theory does not, as is misunderstood by some, claim that genetic drift replaces natural selection in phenotypic evolution. Wright has repeatedly emphasized this point.

Controversy Between Fisher and Wright Wright's evolutionary theory received intense opposition from Fisher's and E. B. Ford's group in England, and a controversy ensued. The issue was whether genetic drift could play an important role in actual evolutionary processes.

According to Fisher, because the number of individuals in a population is extremely large in the majority of biological species, it is highly unlikely that the absolute value of the selection coefficient, which expresses the degree of advantage or disadvantage in selection, will be smaller than the reciprocal of the population size for any mutant gene. Therefore, a mutation that is neutral under natural selection almost never exists, and the force of selection always completely overwhelms the force of genetic drift. On the other hand, according to Wright, a species comprises many regional subpopulations, and as his shifting balance theory shows, genetic drift within a subpopulation plays an important role in adaptive evolution, in particular in the macroevolution of a species. Moreover, according to Wright, in a biological species that forms one large random mating population, the genetic makeup of the species would soon reach an adaptive peak (i.e., a locally optimal state) and no further progress would be possible. Against this argument of Wright's, Fisher believed that such an outcome was not likely. According to the latter, a large population would contain an enormous number of genotypes, and it was claimed that if one of these is adaptive at any one locality, it would increase in frequency and the population fitness would increase.

This controversy between Fisher's group in England and Wright, which was vigorously carried on in the 1950s, was very stimulating to the researchers of the world in the field of evolutionary mechanism theory. I was one of those deeply interested in the course of this controversy at that time. However judging from the reaction of the academic circles of the world around that time, it appears that Fisher's side won.

Wright's shifting balance theory is recently receiving much attention in the field of evolutionary theory. However, it is my impression that although this theory is attractive, there is currently little clear support for it. Of course from 1931 onward, Wright published many extremely important mathematical researches on the chance fluctuation of gene frequencies in finite populations. Without this precedent, it would have been quite impossible for me to later develop the theory of diffusion models (a method that uses a type of partial differential equation) in population genetics.

2.3 Synthetic Theory of Evolution and Panselectionism

Muller's Contribution In addition to the theoretical researches in population genetics carried out by Fisher, Haldane, and Wright in the 1930s, we should not forget the contributions of H. J. Muller in uniting Mendelian genetics and Darwin's evolutionary theory. It is well known that he received the Nobel Prize for his research on the induction of mutations by X-rays, but the magnitude of his contributions to evolutionary genetics is not as well known. Nevertheless, using the fruit fly as experimental material he had, by the early part of 1920s, elucidated the nature of gene mutations and clarified its relation to natural selection, which is indeed a great achievement. In addition, the method first developed by him for detecting mutations induced by X-rays, in particular recessive lethals, is also important. This is a method in which a special strain is constructed that carries chromosomal aberrations such as inversions (a part of the chromosome is inverted) or dominant marker genes and is used for mating; such strains have become indispensable for the analysis of genetic variation in natural populations.

In general, mutational changes, whether they are induced by X-rays or occur spontaneously, are much more often ones that lower the viability of an individual (are deleterious) than raise it (are selectively advantageous). Among them, the experimentally important ones are "recessive lethal genes", that is, mutant genes that cause the death of the individual during development only when they occur in the homozygous condition. In addition it has been found that many "slightly deleterious genes", which lower the viability of an individual by a small amount when homozygous, are also present in natural populations, and the method developed by Muller (or improvements thereof) are even now indispensable for their detection.

Development of the Synthetic Theory of Evolution On these foundations various noteworthy researches in evolutionary genetics eventually blossomed. Especially famous are the genetic analyses of natural populations of fruit flies by T. Dobzhansky and his school; the work of G. G. Simpson who interpreted paleontological data from the standpoint of population genetics and showed that actual evolutionary processes could be satisfactorily explained with reference to genetics; the proposal and development of the "ecological genetics" of E. B. Ford and his school; and Ernst Mayr, an animal taxonomist, who discussed the problem of the origin of species based on distributional and ecological data; such researches were published one after another, giving the impression to many biologists that the synthetic theory of evolution had advanced enormously.

Dobzhansky The focus of Dobzhansky's research was the analysis of the polymorphisms of chromosomal inversions in natural populations. Here, "polymorphism" denotes the phenomenon in which different alleles or different types of homologous chromosomes coexist at high frequencies in the same population. Dobzhansky studied a population in which the so-called inversion chromosomes, where a part of the chromosome is inverted, and the homologous normal type of chromosome coexist, and the discovery that their relative frequencies changed seasonally attracted the attention of many biologists at that time.

In addition, Dobzhansky exerted a strong influence on evolutionary circles by his many excellent publications on evolutionary studies. In particular, his book *Genetics and the Origin of Species* (1937) is famous, and was widely read and highly regarded by biologists. I also read the pirate edition of this book when I was a student after the end of the war, and learned for the first time of the existence of Wright's mathematical studies.

Moreover, Dobzhansky pointed out that overdominance (heterozygotes are more advantageous than homozygotes) played a major role in the maintenance of genetic variation within a species, and gained many supporters. In 1955, Dobzhansky postulated two opposing standpoints in population genetics and named them the "classical hypothesis" and the "balance hypothesis".

According to his definition, the classical hypothesis adopts the position that individuals in the population are homozygous for the wild type gene at the majority of genetic loci, and also supposes that during evolution mutant genes that are advantageous for survival replace the extant genes one after another at each genetic locus. Furthermore, according to this hypothesis, the cases where two or more alleles coexist in the population—and as a result an individual becomes heterozygous at some locus—are when a deleterious gene arises by mutation at a certain rate every generation but is checked by selection and maintained at a low frequency, or when on rare occasions an advantageous mutant gene appears and is found in the transient state of spreading through the population; there are few such cases, and their contribution to heterozygosity is in each case small.

On the other hand, according to the balance hypothesis, the standard adaptive state of an individual is when the majority of genetic loci are in the heterozygous condition; the homozygous condition is disadvantageous in the usual bisexually reproducing population and hence is rarely seen. And because at many genetic loci the existence of multiple alleles (three or more alleles) is favored, it is claimed that many multiple allelic states develop.

Judging from the circumstances at that time, it is clear that Dobzhansky regarded himself as the representative of the balance hypothesis side, and thought of Muller as being the representative of the opposing classical hypothesis side. He of course believed that the balance hypothesis was correct.

This claim of Dobzhansky's subsequently had a great influence on researchers of population genetics in the United States. As seen from the present, it is known that many of the experimental results that he invoked in support of the balance hypothesis were variously problematic. In some cases, a phenomenon that he believed to have discovered (for example, "synthetic lethals") was shown by subsequent detailed research to be almost nonexistent.

I will avoid going deeper into these problems, but there is one thing I would like to mention here. It is about the mechanism whereby recessive deleterious "lethal", "semi-lethal", and "slightly deleterious" genes contained in the population are maintained. Dobzhansky asserted from the standpoint of the balance hypothesis that they are maintained in the population by heterozygote advantage (i.e., overdominance"). However, it was later confirmed by the careful large scale studies conducted by J. F. Crow and his colleagues at University of Wisconsin and by Professor Terumi Mukai and his collaborators at Kyushu University in Japan, that the explanation that they are maintained in the population in the balance between their appearance by mutation and their removal by selection (classical hypothesis), as advocated by Muller from the very beginning, was basically correct.

Exaggerated Advertisement of the Synthetic Theory In England, E. B. Ford and his school forcefully claimed that genetic polymorphism was mainly caused by overdominance. The ecological genetics that he proposed continued to be promoted by his students. In the mid 1960s, a group of extreme panselectionists was born that explained the abundant enzyme polymorphisms discovered in biological populations, i.e., the polymorphisms of enzyme producing genes, by balancing selection, i.e., selection actively working to maintain polymorphism.

In the United States, Ernst Mayr achieved prominence as one of the champions of overdominance and epistasis. He criticized the classical population genetics formulated by Fisher, Haldane, and Wright as inappropriate, and emphasized that the fitness of each gene was a relative one determined by its interaction with other genes. He called this view "the theory of relativity in the field of population genetics". Furthermore, at the Cold Spring Harbor Symposium held in 1959 to commemorate the centenary of the publication of Darwin's *The Origin of Species*, he delivered a keynote address casting doubt on the value of the mathematical population genetics of Fisher, Wright, and Haldane, and called his position the new relativistic population genetics. However, in my opinion, it is not at all like the relativity theory of physics and of little substance.

In this way, the exaggerated advertisement of the synthetic theory of evolution was vigorously pursued, but in retrospect, this was a scientifically sterile period in which there was little fundamental progress. It is regrettable that the tendency to not give due credit to diligent research in population genetics, in particular to underestimate the importance of mathematical research, and to hold that evolution can be understood by flourishing various terms and concepts, was imported into Japan, and even now is fairly influential.

Panselectionism Becomes the Main Stream In any case, in the first half of 1960s, a synthetic theory of evolution bordering on panselectionism came to be accepted as established theory by biologists. The position that regards various traits of biological organisms as products of adaptive evolution was main stream, and the view that neutral alleles that are neither good nor bad under natural selection rarely occur was overwhelmingly held among population geneticists and evolutionists. Around this time Mayr wrote in his tome *Animal Species and Evolution* (1963) that selectively neutral mutants are unlikely to occur, and that it was advisable to stop using genetic drift in explanations of evolution. At about the same time, Ford also wrote in his book *Ecological Genetics* (1964) that neutral mutant genes were extremely rare and could not reach high frequencies in a population. His position was the Fisherian panselectionist one.

Of course panselectionism was not a new idea; it was popular from the end of the previous century to the beginning of this century, but there were so many nonsensical claims that it lowered the reputation of Darwin's natural selection theory. By comparison, the panselectionism that characterized the synthetic theory of evolution of the 1960s incorporated population genetics theory and was scientifically far more advanced.

In retrospect, it was a time when the synthetic theory of evolution and the panselectionist ideas at its core were believed to be unmovable truth. Few researchers in population genetics and evolutionary genetics entertained any doubts, and under its influence my own views were also close to panselectionism. However, it is lucky for me that the diffusion model method, which treats gene frequency change as a stochastic process and on which my main research efforts were focused around that time, later unexpectedly proved to be useful in dealing with variation and evolution at the molecular level from the standpoint of population genetics.

Using this method it is possible to treat the process by which gene frequencies change in finite populations ruled by chance (stochastic process), including the effects of mutation and natural selection. For example, by this method it became possible for the first time to obtain the probability, under general conditions, that one mutant gene arising in a finite population would spread through the entire population (fixation probability), and to compute the conditional (i.e., excluding the cases of loss) mean time required for fixation of the mutant.

2.4 Studies of Molecular Evolution and the Neutral Theory

Two Advances In this way, already in the 1950s, a sophisticated mathematical theory existed for the treatment of gene frequency change in a population as a stochastic process, but it was not possible to apply this to actual problems in evolution and intraspecific variation until research at the molecular level had made head way. The reason for this is that research on evolution and variation had until then dealt with phenotypes (mainly visible morphology) that are far removed from genes, and it was not possible to approach the problem at the level of the internal structure of genes. Therefore, it was not known at what speed new mutant genes turned over within a species during the actual evolutionary process.

However, these limitations were removed when the concepts and methods of molecular genetics were introduced into research on evolution and population genetics from about the middle of 1960s. As a result, two striking advances were made. First, it became possible to compare homologous proteins, specifically the hemoglobin molecule, among vertebrate animals; and, by incorporating knowledge from paleontology, to estimate the amino acid substitution rate during the evolutionary process. The second advance was that it became possible to investigate genetic polymorphism of enzyme proteins in a population by electrophoresis. By this method the first reliable estimates were obtained of variation at the genetic level existing in the species.

Birth of the Neutral Theory The emergence of data on evolution and variation at the molecular level opened the door to a new era in this field. First, with regard to evolution, the amino acid sequences that could be compared at that time were limited to a very small number of proteins such as hemoglobin and cytochrome c; but when I extrapolated from this data to the substitution rate per genome (the haploid set of chromosomes) of mammals, I obtained the surprising estimate that mammalian species had accumulated new mutations (i.e., changes in the DNA bases had been substituted in the species) at a rate of one every 2 years during the evolutionary process. Next, with regard to intraspecific variation, the data at that time on intraspecific variation of enzyme proteins obtained by electrophoresis yielded the result that, in populations of the human and the fruit fly, each individual was in the heterozygous state at more than 1000 loci. This is much greater genetic variation than previously believed.

In order to explain these unexpected results from the standpoint of population genetics, the conclusion I reached in 1967 was that it was necessary to postulate an important role for random drift of mutations, neutral under natural selection, in evolution at the molecular level. After reporting this idea (the neutral theory) at the Genetics Club Meeting in Fukuoka in autumn, I prepared a brief communication and submitted it to the English scientific journal *Nature* around the end of the year; fortunately it was accepted and published in February of 1968.

As I will discuss the neutral theory in detail in Chaps. 7 and 8, I will here briefly dwell mainly on its historical aspects. The neutral theory certainly does not reject Darwin's natural selection theory, but it did not conform to the synthetic theory of evolution that was main stream at that time and thus provoked much controversy. In particular, in the year after my paper appeared in Nature, the publication by J. L. King and T. H. Jukes in the United States of a paper entitled *Non-Darwinian evolution* based on much data from molecular biology—this was essentially the same hypothesis as the neutral theory—stimulated many geneticists and researchers of evolutionary theory, and the debate became even more heated.

Development of the Neutral Theory In this way the neutral theory was for me something that I proposed, compelled by theoretical necessity as entailed by an analysis of observed data; however, as one brainwashed by the synthetic theory of evolution, it was emotionally difficult for me to accept the neutral theory that I had put forward. Around that time experimental results claiming to refute the neutral theory were published one after another in the field of population genetics, but some experiments were faulty, others were incomplete and indecisive; and I remember feeling relieved to have survived the critical 2–3 years following publication.

However, encouraged by the gradual increase of data on molecular evolution that on the whole seemed to support the neutral theory, I continued the work of developing the neutral theory with the cooperation of Dr. Tomoko Ohta.

I arrived at the view around 1973 that amino acid and DNA base substitutions occur more rapidly during the evolutionary process in molecules that are functionally not important (or in unimportant parts of a molecule) than in those that are, and the maximum of the substitution rate (evolutionary rate) is determined by the mutation rate, and was fully aware that this was a totally heretical idea from the standpoint of traditional evolutionary genetics.

Subsequently, after a few years, the second revolutionary age of molecular biology which F. Crick called the "mini-revolution" arrived. And data on the base sequences of DNA began to be published with explosive vigor. As a result, research on molecular evolution apparently shifted from an age in which amino acid sequences of proteins were compared to an age in which base sequences were compared.

From these studies, it also became clear that in DNA base substitutions during evolution, those not causing a change in an amino acid of a protein were occurring at a much a higher rate than those causing a change. Proteins have a fundamental role in forming the body of an organism and in maintaining life; although this function depends on the three dimensional structure, if we consider that it is ultimately determined by the amino acid sequence, DNA base substitutions causing an amino acid change are generally expected to have a much larger effect on the phenotype than those not causing a change. On the other hand, because natural selection acts on the phenotype of an individual and is determined by viability and fertility of an individual, DNA base changes that do not cause an amino acid change are obviously less likely to be subject to natural selection.

Nevertheless, those not causing an amino acid change are substituted at a much higher rate than those causing a change during evolutionary process, and are being accumulated at a high rate in the species.

Establishment of the Neutral Theory In 1977 I made the claim that if the upper limit of the evolutionary rate was shown by future research on molecular evolution to be determined by the mutation rate, it would support the neutral theory; after a while, dramatic data in its support appeared from an unexpected direction. This was with regard to the hemoglobin pseudogene discovered in mice. Here, a pseudogene is a gene that is quite similar to a known normal gene in its base sequence, but which for some reason has lost its function as a gene after arising by duplication from a normal gene, and is sometimes called a "dead gene". Dr. Takashi Miyata of Kyushu University succeeded in estimating its evolutionary rate for the first time in the world. The value obtained was apparently the largest for a DNA base substitution rate. A pseudogene is originally derived from a normal gene and for some reason has lost its function as a gene. Therefore, it has no phenotypic effect, and whatever mutations occur have no deleterious effects; hence the observed result can be most readily understood if a pseudogene is regarded as not being subject to natural selection (has become neutral) and to be undergoing evolutionary change at the maximum rate set by the neutral theory.

In general, when homologous base sequences are compared, it has recently become common practice to predict an important function for parts that do not change in evolution, and to regard parts that change rapidly as being functionally unimportant. However not many people seem to realize that this point of view is based on the neutral theory. I cannot deny that I feel secretly happy that this point of view, derived from the neutral theory, has become common sense in research on molecular evolution.

A big remaining question is whether evolution at the phenotypic level can be linked to evolution at the molecular level. I hope that achievements in this area gaining international recognition will be made by young Japanese researchers in the future.

The history of science tells us that, when a new research area is pioneered, new phenomena are encountered that cannot be explained by extant concepts, and new ideas are required for dealing with them. In the history of research on biological evolution I would like to think that the neutral theory has in some small measure played such a role.

2.5 Other Evolutionary Theories

Punctuated Equilibrium Theory Lastly, I would like discuss two theories which, although not directly relevant to evolutionary genetics, are recently receiving world-wide attention in the field of evolutionary theory.

First, there is the "punctuated equilibrium theory" proposed by N. Eldredge and S. Gould at the beginning of the 1970s. The essence of their claim is that evolution does not proceed gradually at a constant rate, but rather consists of brief periods of rapid change and long intervals of stasis in between. This theory was put forward in clear opposition to the traditional view, which holds that evolution proceeds by each lineage of organisms slowly accumulating small variations, and achieved instant fame due partially to being advertised in the attractive and eloquent publications of one of its proponents, Gould.

Of course, evolution of the phenotype (mainly morphology) is being considered here. According to punctuated equilibrium theory, periods of abrupt change coincide with periods of speciation. In addition, the concept of "species selection" is emphasized in the claims of the punctuated equilibrium theorists. In other words, it is claimed that species with special attributes have an advantage in interspecies competition over species lacking them, are more likely to give rise to descendent species, and anagenesis (evolutionary change that can be regarded as progress, including organizational complexity) ensues. In fact, according to Gould, the two traditional concepts of anagenesis and cladogenesis—branching of descendants—are not independent, but rather anagenesis is the result of interspecies competition acting on species divergence.

This theory of Gould and others is not accorded that much importance by geneticists. This may be because there is much data showing that morphological change and the development of reproductive isolation have genetically different bases, and moreover, conceptually speaking, there is little novelty compared to Wright's shifting balance theory which was proposed 40 years earlier. Nevertheless, punctuated equilibrium theory is not without importance in the field of evolutionary theory as it greatly stimulated researchers in paleontology.

Symbiosis Theory Second, there is the "symbiosis theory" on the origin of the eukaryote cell, vigorously proposed and developed by the American scientist, Lynn Margulis, from the 1970s onward. According to her theory, organelles such as mitochondria, chloroplasts, and flagella in the cells of eukaryotes are derived from ancient freely-living microorganisms (prokaryotes) that entered these cells and became symbiotic. Here, by eukaryotes we mean organisms with a clear nuclear structure inside the cell surrounded by a nuclear membrane; animals and plants with which we are familiar belong to this group. By contrast, prokaryotes are a group represented by bacteria and cyanobacteria, which are primitive organisms without such a nuclear structure within the cell. The two differ strikingly not only with regard to the nucleus, but also in many other traits, and the biological world can be partitioned into these taxonomical groups.

According to Margulis, mitochondria are descended from ancient aerobic bacteria, whereas chloroplasts arose from cyanobacteria and flagella from spirochetes which underwent changes after forming symbiotic relationships with ancestral eukaryote cells. The idea that chloroplasts and mitochondria arose from symbiosis was not new, and had been entertained by some biologists in a vague form since the discovery of mitochondria, etc. However, Margulis' achievement was that she integrated new discoveries in biology and made this theory incomparably more substantial than before. Particularly noteworthy is that it has become clear from research in molecular biology that mitochondria and chloroplasts each have distinct genetic material (DNA), with characteristics close to the DNA of prokaryotes. The hypothesis that cellular organelles of eukaryotes are not the result of changes inside the cell (for example, invagination of the cellular membrane), but are the result of parasitization by or symbiosis of prokaryotes will provide an important basis for thinking about the evolution of primitive cells.

3 Tracing the Course of Evolution

3.1 Outline of the History of Life

Birth of Life Let us first take a brief look at the history of life on the Earth. It is estimated that the Earth on which we live was born as a member of the solar system approximately 4.6 billion years ago. And it is thought that the first life on the Earth, that is self-replicating molecules resembling genes, were born several 100 million years later, about 4 billion years ago.

Around that time the crust was formed and the oceans were in existence. The seas are the cradle of life, and the currently most plausible explanation is that the first life was formed in the seas from lifeless organic matter according to the laws of physics and chemistry.

The biggest difference in the terrestrial environments then and now is the composition of the atmosphere, there being almost no free oxygen in the atmosphere around that time. As a result, there was no ozone layer high in the sky, and strong ultraviolet radiation from the Sun must have directly impinged on the surface of the Earth. This provided one important energy source for the chemical reactions necessary for the origin of life. Then, once life was born, new organisms subsequently arose one after another by biological evolution, eventually giving rise to the enormous diversity of the biological world witnessed today. Because the 4.6 billion years since the birth of the Earth is much too long compared to the time scales experienced in our daily lives and is hard to imagine, Fig. 3.1 compresses this into one year and furthermore subdivides this into 12 months to show the major events that took place on the Earth. In this representation, one month corresponds to a little less than 400 million years.

Ancient Microorganisms and the Accumulation of Oxygen If we express time in this way, the Earth of course began on January 1st, and the origin of life can be dated to about the middle of February. The oldest rocks now known were found in Greenland and are estimated to be 3.7 billion years old, which corresponds to March. The oldest fossils of microorganisms are at 3.2 billion years ago, and this

M. Kimura, *My Thoughts on Biological Evolution*, Evolutionary Studies,
https://doi.org/10.1007/978-981-15-6165-8_3

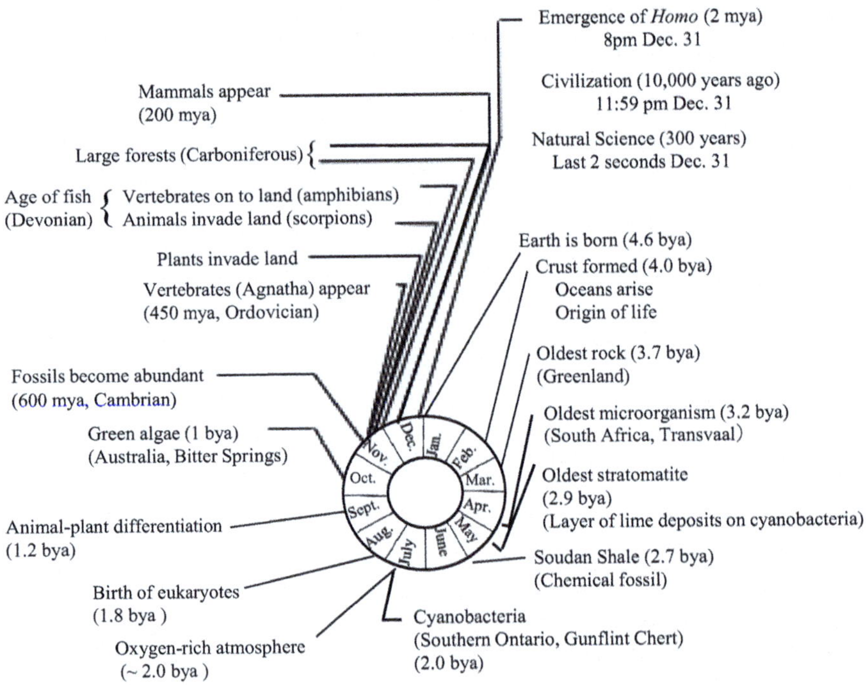

Fig. 3.1 History of the Earth reduced to one year (*bya* billion years ago, *mya* million years ago)

is already the last third of April. Furthermore, sediments composed of cyanobacteria and limestone called stromatolites are known, the oldest of these being 2.9 billion years old, which is the middle of May. Interestingly, stromatolites are even now actually being formed in Australia and South Africa. Free oxygen gradually increased in the atmosphere due to the photosynthetic activity of cyanobacteria, and by approximately 2 billion years ago a stable oxygen-rich atmosphere had emerged, and this is the last third of July.

It is absolutely impossible for us now to survive without oxygen. However, oxygen was highly poisonous for early organisms. The organisms at that time had evolved under oxygen-free conditions and were almost all anaerobic. Hence, when oxygen came to be produced by cyanobacteria and began to accumulate in the atmosphere, this posed a big threat to the survival of the majority of organisms. What saved the day was symbiosis, which was explained in the section on Margulis' theory in the previous chapter. Namely, these organisms happened to incorporate microorganisms into their cells with the ability to utilize oxygen and were able to overcome the crisis with their help. Mitochondria which all higher organisms now possess as cellular organelles are thought to have arisen in this way. And because the atmosphere has been rich in oxygen throughout the past two billion years, and because they have evolved for a long time under such conditions and have completely adapted to such an environment, organisms like those we see around us have changed so that they cannot survive without oxygen.

Appearance of Eukaryotes and Divergence of Animals and Plants About 1.8 billion years ago (August in the figure), eukaryotes were at last born. Until then, there were only lower organisms such as bacteria and cyanobacteria, which did not have a nucleus within the cell, but around this time organisms possessing a nucleus appear. The emergence of a nucleus entails that chromosomes are formed inside it, on which many genes come to reside. With many genes it became possible to store complex genetic instructions, and as a result the foundations were laid for the birth of higher organisms comprising many cells and with complex systems. As to how the nucleus arose, there is at present no convincing explanation.

The common ancestor of animals and plants is thought to have appeared about 1.2 billion years ago. This corresponds to the last third of September. This estimate, as well as the estimated age for the birth of eukaryotes, was obtained by molecular phylogenetic methods to be described in Chap. 8. In this way, theoretically speaking, the path to animals and the path to plants diverge 1.2 billion years ago, but the remarkable evolution of plants on land began much later, 800 million years later, just before the Devonian Period (the last day of November in the figure).

Toward Multicellular Organisms It is in the Cambrian Period that the actual history of evolution becomes clear. This is the period that began about 600 million years ago and continued until 500 million years ago, and the beginning of the Cambrian already corresponds to the middle of November in our calendar. From this period fossils of many animals with advanced structural plans are found, such as trilobites. Many of these organisms had hard body parts, and were more likely to endure as fossils.

The Cambrian Period and the several hundred million years that preceded it were a time during which there was extremely active evolution toward a variety of higher multicellular organisms, and the seas around that time were, so to speak, one big experimental arena for the evolution of higher organisms. Seeing that almost all the major groups of animals present today (specifically "phyla" in the terminology of taxonomy)—but not plants—had appeared during the Cambrian Period and the subsequent Ordovician (about 60 million years in length), we can infer that this was a time when remarkable adaptive radiation occurred. In general, multicellular higher organisms have the potential to adapt to a much greater variety of environments than the antecedent lower unicellular organisms. Thus, during this time, various structural plans were tried in the lifestyles newly available to the multicellular organisms.

In fact, among the fossils of the Cambrian Period we can see animals with very peculiar structural plans that do not belong to any of the current groups of animals, and these can be regarded as failures in the evolutionary trial-and-error that dropped out.

Advance on to Land The direct ancestor of vertebrates like us is a primitive fish without a lower jaw, which appeared between the end of the Cambrian Period and the Ordovician Period. It is in the Devonian Period that various fish appeared, which is between 350 million and 400 million years ago. Transformed to one year, this is around the beginning of December. I will discuss the evolution of vertebrates in detail in the next section and will omit it here, but I would like to briefly consider the advance on to land of organisms. Since the birth of life, the evolution of organisms occurred for a long time in the oceans, but plants began their invasion of land just before the Devonian. The first terrestrial plants are thought to be Psilotopsida. Animals followed plants on to land during the Devonian Period, and amphibians and wingless insects appeared around this time.

Age of Dinosaurs In the following Carboniferous Period (between about 360 million and 290 million years ago), ferns and conifers formed large forests; reptiles began evolving from around the end of this period. During the Mesozoic Era which covers the Triassic (between about 250 million and 210 million years ago), Jurassic (between about 210 million and 140 million years ago), and Cretaceous (between about 140 million and 65 million years ago) Periods, reptiles and in particular dinosaurs proliferated, and it can be called the age of dinosaurs (the Mesozoic Era corresponds to about December 11–26). In addition, this was the period in which ammonites flourished in the seas, and on land plants such as ferns, ginkgo, and cycads grew thick.

Mammals of which *Homo* is a member appeared about 200 million years ago, which is the middle of December. Mammals during the Mesozoic Era are believed to have hidden from dinosaurs, surviving as small nocturnal organisms. Moreover, flowering plants developed in the second half of the Mesozoic Era, and the remarkable evolution of insects occurred in parallel. This demonstrates the intimate existential relationship between insects as pollinators and plants. The "coevolution" of insects and flowering plants is an extremely interesting problem, and I would like to add that Dr. Peter Raven, the second recipient of the International Prize for Biology newly established in Japan, has done excellent research on this.

Birds arose from one lineage of dinosaurs during the Mesozoic Era and gradually developed from about the middle of this era.

Age of Mammals, Age of *Homo* About 65 million years ago dinosaurs and ammonites suddenly perished, and a major change occurred in the fauna and flora on the Earth. This marks the end of the Mesozoic Era and the beginning of the Cenozoic Era. In the scale converted to one year, there are about 5 days remaining in December. During the Cenozoic Era mammals progressed and developed remarkably, and among plants ferns and conifers declined and in their place flowering plants developed strikingly. Birds also underwent remarkable advance. The major part of the Cenozoic Era is called the Tertiary Period, and the following Quaternary Period beginning about 1.8 million years ago is, as it were, the age of *Homo*. *Homo* is thought to have been born about 2 million years ago, which correspond to 8 pm on December 31.

Homo developed civilization with farming and writing from 10,000 years ago; this corresponds to 11:59 pm of December 31st, and in terms of the scale in which the history of the Earth is compressed into one year, it can be said that *Homo* acquired civilization and at long last achieved a human lifestyle just before the last minute. About 300 years ago, humans used their intellect to create natural science. In other words, they began trying to objectively understand all natural phenomena; in the scale converted to one year this occurred in the last 2 s, a very short time ago.

3.2 Evolution of Vertebrates

Thanks to Sea Scorpions Let us now look at the evolutionary history of the vertebrates to which we belong. In the previous section, I noted that fossils provide a clear understanding of the course of evolution from the Cambrian Period onward; near the end of this period, about 500 million years ago, our distant ancestor and the oldest true vertebrate, the Agnatha, appeared. These are primitive types of fishes without a lower jaw, which lived by swallowing mud from the sea bottom with their mouths and extracting food from it. In other words, we can imagine that they led a life similar to the worm. The size of their bodies was apparently between several centimeters and 50 cm in length and was generally small. And they had a curious appearance with unpaired fins. Their fossils are often seen in the Silurian Period (between about 440 million years ago and 400 million years ago); their bodies are covered with a carapace like a suit of armor, and they are also called ostracoderms. This suit of armor is believed to have developed as protection against a predatory arthropod called the sea scorpion (eurypterid) which was dominant at that time.

In fact, from the end of the Silurian Period to the beginning of the Devonian (about 400 million years ago), the fauna as judged from the occurrence of fossils appears in many cases to consist mainly of small Agnatha and large sea scorpions. The sea scorpions at that time were ferocious carnivores, some of them reaching 2.5 m in length. And we can imagine that they lived by preying on Agnatha. When, among the descendants of Agnatha, bony fishes with a developed structural plan that were good swimmers eventually appeared and began to thrive, the sea scorpions decreased in numbers and then went extinct by the end of the Paleozoic.

In this way, the bones of vertebrates first came into existence as plates covering the body surface, then eventually bones were formed inside the head and continued to evolve into a back bone supporting the body like a shaft. In retrospect, vertebrates with further improvements could not have emerged without a skeleton, and this was originally armor for protection against enemies. In particular, it arose because there was a fearsome enemy like the sea scorpion; the famous American paleontologist A. Romer once said that we must thank the sea scorpion, which was the enemy of our ancestors of ancient times. This event reminds us of the proverb "adversity makes a man wise".

In connection with this, although not directly relevant to the evolution of vertebrates, a problem that comes to mind is the large difference in structural plan between Pelecypoda (bivalves) and Cephalopoda (octopuses and squids), although they are both members of Mollusca. Namely, whereas octopuses are a "higher" advanced form with the highest intelligence among invertebrates, bivalves are a degenerate form entirely lacking a head. Bivalves were likely well protected by a hard shell, and the great success they enjoyed by their sessile life style may have had the perverse effect of blocking the path to progress. The warning "do not withdraw into your shell" may be applicable, not only to our human society, but to biological evolution as well.

Humans Might Have Been a Thousand-Armed Kannon (Avalokiteshwara) Returning to the main argument, in the later part of the Silurian Period (between about 440 million years and 400 million years ago) which comes after the Ordovician, a pair of bones that used to support gill openings were enlarged, jaws developed, and a new age dawned in the evolution of fishes. In the Devonian (between 400 million and 350 million years ago) jawed fishes underwent major developments and radiation, and at the same time jawless ostracoderms declined and were almost extinct by the end of the Devonian. Only the lampreys and hagfishes survive to this day as degenerate forms. During the Devonian Period, there were many kinds of fishes, but one of the major types with jaws were the placoderms. Placoderms are, as this name indicates, fishes with the body surface covered by a plate-like hard shell, and this type flourished but were also almost extinct by the end of the Devonian.

These placoderm fish had very strange appearances compared to the present day fish; for example *Bothriolepis* was shaped exactly like a cross between a turtle and a shrimp. The scholar who first discovered this fossil was for a long time under the impression that this was the fossil of a turtle. However, later research showed that this was after all a type of fish. Arms like those of a shrimp extend from the size of the body, and it is thought that these appendages were used to obtain food and to walk on the bottom of the water.

In addition there were the ancestors of today's sharks and other bony fishes, and among those thought to group with the latter were fish with many side fins. If they had come onto land and become our distant ancestor, the structure of our bodies would have been quite different from what it is now. In other words, as J. B. S. Haldane once said, our bodies might not have been shaped with two arms and two legs, but might have been something with many arms like a Thousand-armed Kannon (Avalokiteshwara).

Conquest of Land The next big step in the evolution of vertebrates was the conquest of land, which began with the amphibians and was completed by the reptiles. According to geological evidence many shallow swamps existed during the Devonian Period, and seasonal droughts often occurred. Under such conditions, fish with the ability to crawl out of dried-up swamps and move to other water-filled ponds with the help of their fins undoubtedly had more opportunities to survive. In this way fins at last developed into feet. Moreover there were frequent droughts, and it is likely that being able to breath with the air bladder was advantageous for survival, and this also triggered the evolution of living on land. Eventually a group with fleshy leaf-like fins called the crossopterygians gave rise to the amphibians, which became the ancestor of terrestrial vertebrates. The crossopterygians prospered during the Devonian, but rapidly declined at the end of the Pliocene. They apparently lost in competition with the amphibians which are their descendants and lived under similar environmental conditions.

As mentioned in the previous section, terrestrial plants grew thick forming large forests on the Earth during the Carboniferous Period, and in association with this insects were enjoying great success by the end of Carboniferous. These insects are believed to have been an important food source for the reptiles, enabling the latter to undergo large scale radiation during the Permian Period, the final period of Paleozoic, and the following Mesozoic.

In the second half of the Carboniferous Period about 310 million years ago, reptiles evolved from amphibians. One important characteristic of reptiles that distinguished them from amphibians was the presence inside the egg of a thin sac called the "amnion", such that the offspring developing from a fertilized egg was protected by amniotic fluid during growth. Because of this (and of course with other changes) reptiles were liberated from the necessity of returning to water for reproduction as in the case of amphibians, and were able to move freely on land.

The fact that in the process of the invasion of land by multicellular organisms, three types of organisms—plants, insects, and terrestrial vertebrates—emerged and underwent major developments over a span of 100–200 million years demonstrates how effective the emergence of a new unexploited living space is in promoting large scale evolution. Moreover, the atmosphere around that time contained about the same amount of oxygen as now, and as a result there was an ozone layer high in the sky, which protected the terrestrial organisms from the lethal effects of ultraviolet radiation. Without this it would have been impossible for higher animals to live on land.

Pangea In the terminal Paleozoic, in contrast to the situation now, the continents existed in one mass as a super continent called Pangea. Later, this continent gradually split up and the fragments drifted apart, and eventually the present continents were formed. This "continental drift" hypothesis was first proposed more than half a century ago by the German geophysicist Alfred Wegener, but this hypothesis was greeted with scorn and disbelief. A little later Wegener met with disaster and disappeared while exploring Greenland; his theory has been shown to be essentially correct by recent research in the earth sciences. Here again we see the tragedy of a great pioneer.

Coelacanth It was believed by paleontologists for a long time that the crossopterygians, which gave rise to the amphibians, had gone completely extinct. However 20–30 years ago, a relict species belonging to this group called the coelacanth was caught in the deep seas of the Indian Ocean and named *Latimeria*, which was big news in science. Recently, it has been found that there are quite a number of surviving coelacanth individuals, and it has been reported that a Japanese expedition obtained a specimen. In any case, the coelacanth is the most famous example of a "living fossil".

The Mystery of Mass Extinctions One other event that I would like to mention in connection with the Devonian Period is the large scale extinction of organisms which occurred 370 million years ago. At that time an unimaginably large tsunami occurred, the reefs made by organisms in the seas of the world were completely destroyed, and the organisms living on the seashores and nearby places also suffered heavily. However, the fishes in the seas suffered relatively little harm. It is believed that the cause of the tsunami was not an earthquake at the bottom of the sea, but rather that it was induced by the impact of something like an enormous meteor or an asteroid striking the sea bottom.

As it happens, this was not the first time that a mass extinction of biological species occurred on a global scale in a short time, and it is clear that there were at least two similar events before this. The first was about 500 million years ago at about the end of Cambrian Period, and this resulted in the annihilation of the majority of trilobites. The second was about 440 million years ago around the end of the Ordovician, and it is said that about 57% of the genera of marine invertebrates went extinct. The third mass extinction occurred during Devonian as already mentioned. The fourth occurred around the end of the Permian, which is the last period of the Paleozoic about 240 million year ago. This was on a most tremendous scale seen for the first time in the history of life, and it is said that 96% of the species of marine animals alive at that time went extinct. Trilobites also disappeared completely, and the Pliocene ended. The fifth (last) mass extinction occurred around 65 million years ago; I will deal with this later, but the dinosaurs went completely extinct, the Mesozoic ended, and the Cenozoic began.

Incidents in which biological species underwent mass extinction in a short time were not limited to the five occasions mentioned above; there were a few others although their scale was somewhat smaller. According to D. M. Raup and others, there were a total of 22 mass extinctions from the Cambrian Period onward, occurring with a 26 million year cycle. To explain this, they proposed the hypothesis that the Sun has an unknown companion star that they provisionally named Nemesis, which periodically perturbs the orbits of asteroids and comets, disposing them to strike the Earth, and thus resulting in the mass annihilation of biological species. However, evidence in its support has not yet been obtained.

Success of Reptiles The super continent Pangea was formed toward the end of the Paleozoic Period as mentioned earlier, and during the Mesozoic which comes after the Paleozoic, reptiles were kings of Pangea. It is estimated that reptiles (Amniota) emerged by evolving from one lineage of amphibians (not a very successful lineage at that) about 310 million years ago. This corresponds to the second half of the Carboniferous Period. Reptiles at first were animals that ran unsteadily on land, but subsequently through adapting to catching insects there appeared some that were adept at running on four legs.

The development of jaw muscles was also important, and eventually a reptile called Synapsida evolved, which became the ancestor of mammals. They are also called mammal-like reptiles, but the Latin name Synapsida is derived from the presence in the cranium of a groove for the jaw muscles shaped like a bow. This group of reptiles emerged in the second half of the Carboniferous, flourished until around the end of Permian (about 250 million years ago), and included an extreme diversity of types. They remained rulers of the land until their place was usurped by the dinosaurs in the Mesozoic. In the Mesozoic they lost in competition with the efficiently bipedal and highly predatory dinosaurs.

The Mesozoic is called the age of reptiles, but dinosaurs are not the only ones that flourished, and there were in addition flying reptiles, marine reptiles, and the ancestors of birds. The dominant reptiles (Archosauria) were extinct by the end of Mesozoic, but birds still survive and are flourishing as one class of vertebrates. Birds are elegant, and are protected from terrestrial predators by their energy-expensive life in the air, but in places without predatory animals the ability to fly always degenerates, and there is a tendency to adapt to living on the ground.

3.3 Evolution of Mammals

The First Mammal The first mammal was born about 200 million years ago as the descendant of a mammal-like reptile. The fossil record shows that the mammals of that time were small nocturnal animals like today's mouse, and it is imagined that they lived by avoiding dinosaurs. Also, constant body temperature and fur developed as adaptations to life on cold nights. According to research on fossils, the early mammals of 200 million years ago had bodies that were only about one tenth the size of the ancestral mammal-like reptiles, but the relative size of the brain was 4 to 5 times as large. There was a need for smell and hearing to develop for life on dark nights, and it is thought that enlargement of the brain was brought about because of this (Fig. 3.2).

For mammals the Mesozoic was a time of trial but also meaningful. This is because the superior properties of mammals, in particular intelligence and the ability to function independently of temperature, were vastly improved in order to survive under the reptilian tyranny. The paleontologist A. Romer in touching on this problem has said that, as mammals, we owe a debt of gratitude to the dinosaurs for their unintended aid.

Mammals have the characteristic of raising their children and taking good care of them. By this, the time required for the development of the brain is secured, and the mature brain is used efficiently throughout the life of the individual. The development of a superior cerebrum can be said to be the true cause of the success of mammals.

Did Collision of an Asteroid Cause the Extinction of Dinosaurs? At the end of the Mesozoic Era (about 65 million years ago), the various dinosaurs that had flourished until then all went extinct. Not only that, many other animals and plants including the ammonites suddenly died out at this time. Because this event occurred at the boundary between the Cretaceous and Tertiary (this is usually abbreviated the K/T boundary rather than the C/T), it is called the K-T extinction, but why such a

Fig. 3.2 Reconstruction of the ancestor of mammals (200 million years ago) (from Nature, Vol. 272, 1978)

mass extinction of organisms occurred was for a long time a big mystery in paleontology.

However, L. W. Alvarez and others recently proposed a bold and promising hypothesis. Their hypothesis is based on data showing a sudden increase in the concentration of iridium, which is rare on the Earth but is relatively abundant in meteorites, in the strata of exactly this period. According to their hypothesis, an asteroid (with an estimated diameter of 10 ± 4 km) struck the Earth, and because of this impact a large amount of dust was thrown up and entered the stratosphere. The dust drifted through the air for several years, spreading over the entire planet. Because of this, the surface of the Earth was darkened and photosynthesis by plants declined, destroying the food chains at their origin. And thus the large herbivorous and carnivorous animals including dinosaurs are said to have gone extinct.

It seems that this hypothesis has subsequently been strengthened by supporting observations appearing one after another. It is difficult for us now to believe that such catastrophes are not fantasy but real. But according to researchers of asteroids, the average waiting time for an asteroid to collide with the Earth is proportional to the square of its diameter. For example, an asteroid with a diameter of 10 km apparently collides with the Earth once every 100 million years, one of 1 km once in 1 million years, and one with a diameter of 100 m strikes the Earth once every 10,000 years on average.

Evolution and Chance Mammals did not show much diversity at first, but spectacular radiation began in the Cenozoic, and they filled the various ecological niches previously occupied by the extinct reptiles. In particular the ordinary mammals excluding the marsupials, in other words the placental mammals, were very successful, undergoing astonishing diversification and making the Cenozoic Era "the age of mammals".

In this connection, the asteroid collision theory of Alvarez and others gives us much food for thought. The implication that our existence is from an evolutionary standpoint ruled by chance is especially important. In other words, as N. Calder has repeatedly pointed out, it is conceivable that if this asteroid had crossed the orbit of the Earth just 20 min earlier or later, it would not have collided with the Earth, consequently there would have been no extinction of reptiles, the large scale adaptive radiation of mammals would not have occurred, and therefore hominins would not have appeared.

Spectacular Adaptive Radiation That placental mammals (eutherians) underwent spectacular adaptive radiation in the Cenozoic era is clear from the diversity of mammalian groups that arose as a result. In what follows, I will briefly list the major groups (corresponding to the taxonomical "order").

First, a very successful group that also achieved an improved structural plan are the carnivores, whose members include meat-eating cats, dogs, otters and bears; A. Romer once stated something to the effect that it is extremely disagreeable to our moral sense that they are evolutionarily so successful. Perissodactyla (horses, rhinoceroses, etc.) and Artiodactyla (cows, pigs, etc.) have adapted to an herbivorous life. Proboscidea are a unique group of African origin that are closely related to them, and whose only current survivors are the Indian and African elephants. Cetacea have adapted to a life in the oceans and their hind limbs have degenerated completely. Some whales are 30 m in the length and 200 tons in weight; they are the biggest animals to have inhabited the Earth. Sirenia also live in water, but are phylogenetically close to the Proboscidea. Among the placental mammals, the Insectivora are thought to be primitive; Chiroptera (bats and their kind), Primates (monkeys and their kind), Lagomorpha (rabbits and their kind), and Rhodentia (squirrels, rats, and their kind) are thought to be close to the Insectivora, but each has a distinctive life style.

3.4 Evolution of Primates and the Emergence of Hominins

Primates Among these various placental mammals, the primates include a wide range of forms—from the primitive lemur to the Japanese macaque and apes that are even more advanced—and also the human, making their evolution of great interest to us. Hence, I would like to focus on the emergence of hominins and to briefly discuss their evolution.

With a few exceptions like the human, the primates are generally tree-dwellers, usually live in mild climates, and many are omnivorous with a bias toward herbivory. Their common ancestor is believed to have been an animal resembling the tupai that lives in eastern India today.

Many features which characterize the primates are the products of life in trees. Prominent among them are hands with strong grasping power and opposable thumbs, eyes forwardly turned with well-developed sight, and the ability to gauge distance with great precision. The cerebrum is highly developed in order to control these faculties. However, the sense of smell is much reduced.

Australopithecus According to the fossil evidence from Africa, *Australopithecus* (meaning "southern ape"), which emerged about 4 million years ago or a little earlier, represents the direct ancestor of humans. It is classified as *Australopithecus* which is different from *Homo*, but has properties quite like humans. However they were small compared to the present-day humans (height about 130 cm), and brain volume was about one third. They prospered, but disappeared about 1.3 million years ago.

The evolution of humans took off when they abandoned life in the trees, and an upright posture with bipedal gait was established by some 3 million years ago. It is said that in the late Tertiary the forests dwindled and were replaced by grassy savannas. This forced the ancestors of humans to become ground dwellers. In the words of A. Romer, humans did not leave the trees, but trees left the humans.

Homo habilis As creatures humans are not exceptionally strong, and after leaving the protection of the trees they were only able to survive by the cunning which was made possible by cerebral development. About 2 million years ago *Homo habilis* (*habilis* means "able"), the first member of the genus *Homo*, apparently emerged in eastern Africa and survived for about 500,000 years. They were clearly the descendants of *Australopithecus*, were widely distributed throughout Africa, and used tools (stone tools). The size of the brain was approximately intermediate between that of *Australopithecus* and the modern human.

As the American geneticist H. J. Muller once pointed out, there is likely to be positive feedback between the use of tools and the genetic ability for their skillful use. That is, individuals with such a genetic ability have an advantage in survival and tend to make even better tools, and this induces selection for a new genetic ability for the use of tools. In this way, the intelligence of human ancestors improved rapidly by natural selection.

Homo erectus The next member of the genus *Homo* to emerge was *Homo erectus*, which flourished between about 1.7 million years ago and 500,000 years ago (needless to say, this name derives from upright walking). Until then Africa was the stage for the evolution of our ancestors, but *erectus* dispersed into Asia and Europe.

This species closely resembles today's humans in the size of the body and in other characteristics, and brain volume was fairly close to that of the human. Also evidence for the use of fire has been found at a campsite of this species estimated to be 1.4 million years old. Moreover it is known that various stone tools were used to cut meat and wood, as well as for other purposes.

The Great Ice Age Lastly, let me discuss the evolution of *Homo sapiens*, the species to which we belong. Here "*sapiens*" means wise. The oldest known fossil of this species is about 500,000 years old and was unearthed in Europe. Subsequently, Europe becomes the main stage of human evolution, and therefore I must touch on the climate of Europe at this time.

The last period of the Cenozoic Era, the Quaternary, which begins 1.7 million year ago and continues up to the present, is mostly taken up by the Pleistocene. The Pleistocene Epoch spans the greater part of the Quaternary Period (between 1.7 million years ago and about 10,000 years ago), excluding the Holocene which begins about 10,000 years ago, and is characterized as the Great Ice Age. In other words, the human species is the product of the glaciations. We are particularly interested in the late Pleistocene (between about 700,000 years ago and 10,000 year ago), during which Europe was visited by many glaciations. During the past 550,000 years there were at least six major glaciations, interspersed by relatively short interglacials. The last glaciation began about 70,000 years ago, reaching its coldest phase about 18,000 years ago; the temperature rose after this, the glaciation ended about 10,000 years ago, and we are currently living at a very warm time. According to recent information, past records show that such warm interglacials do not last a long time. Hence, the argument that the next glaciation will soon occur (not more than several tens of thousands of years in the future) seems persuasive.

Neanderthals and Cro-Magnons About 120,000 years ago, hominins called Neanderthals appeared, who are known only from fossils and flourished across Europe and in western Asia for 80,000 years. They developed a unique culture, but disappeared a little over 30,000 years ago. Their name derives from the discovery of the oldest fossil from the valley of the Neander River in Germany. Biologically speaking, they are currently considered to belong to the same species as humans, and have been given the scientific name *Homo sapiens neanderthalensis* as one human subspecies. Their build was thickset compared to the present day humans, their height was relatively short at a little more than 150 cm, their heads were large and cranial capacity was slightly larger than modern humans. There is evidence that they cared for the injured and the old, and buried the dead with weapons and flowers. Generally speaking, the custom of caring for the old may have contributed to the progress of a tribe by facilitating the transfer of knowledge from the old to the young. In other words, it was adaptive for the transmission of culture. It is thought that eventually they were defeated in the struggle for existence by the Cro-Magnons who invaded Europe from southwestern Asia about 40,000 years ago, and perished. However, some interbreeding apparently occurred.

The Cro-Magnons are essentially the same as modern humans, and both are biologically classified in the human subspecies *Homo sapiens sapiens*. Based on research on the cranial bones of Neanderthals, it has recently been claimed that their pronunciation was not as fluent as modern humans, and in particular it seems they were unable to pronounce vowels such as [i], [u], [a] and consonants such as [k] and [g]. Perhaps the development of the parts of brain for the efficient transmission of complex information by the spoken word was vastly inferior to modern humans, and they were defeated in the struggle with the Cro-Magnons for this reason. The Cro-Magnons used advanced stone tools and did a lot of hunting. Lascaux Cave in France is famous for the magnificent pictures of cows and other animals drawn by them. It is estimated that at the end of the Pleistocene, about 10,000 years ago, the total population of the world was about 10 million.

The story of human evolution summarized above is based on the currently prevailing views of researchers in this field. We must not forget that there were many twists and turns, and that the accumulated efforts of many scholars were involved in reaching this understanding.

When Did the Human and Apes Diverge? I would like in passing to touch on the question of when the human and the African apes (chimpanzee and gorilla) diverged.

There has been much discussion on this question in the past. According to traditional paleontological research based on comparisons of anatomical structures (e.g. teeth), the point of divergence was said to be between 20 million and 30 million years ago. In particular, the fossils named *Ramapithecus* and *Sivapithecus* dated to 14 million years ago were thought to represent the direct ancestors of humans, and therefore it seemed certain that the human and apes had diverged before then.

In contrast, the estimate of the divergence time using molecular phylogenetical methods to be described in Chap. 8 is at most 5 million years ago. Of course this result met for a while with vehement opposition from the researchers of fossil hominins, and doubt was expressed toward molecular methods. However, many fossils have recently been obtained, and *Ramapithecus* and *Sivapithecus* which were previously thought to be the ancestors of humans are now believed to be closer to the orangutan. Therefore the fossil record was seen not to be inconsistent with the estimate of at most 4–5 million years for the divergence time of the human, on the one hand, and the chimpanzee and gorilla, on the other. Moreover, the reliability of molecular phylogenetical methods has gradually won wide acceptance, and with regard to this divergence time the result obtained from molecular methods is now thought to be more secure than the estimate using fossils, even among researchers of hominin fossils. This can be said to be a testament of one victory in molecular evolutionary research.

Mutation as an Evolutionary Factor

4

4.1 A Genetic View of Life

Cells, Chromosomes, Genes The bodies of the human and other higher animals and plants are all made of microscopic units called "cells" (or materials produced by them). This claim is based on the "cell theory" proposed by M. J. Schleiden and T. Schwann about 150 years ago, and the substance of this claim is now widely accepted as a self-evident "truth" by biologists in general. Most human cells have a diameter of the order of 10–100 μm. This entails that ordinarily about 100 million cells are required to make a piece of tissue 1 cm^3 in size, and we can see from this how small the cell is in comparison to the scale of our daily lives (the total number of cells in the human body is about 10^{13}).

In principle all cells have a nucleus, and the nucleus contains two sets of chromosomes derived from the two parents. In the human the nucleus of a somatic cell contains 46 (i.e., 23 pairs of) chromosomes, and one among these pairs are the "sex chromosomes" which determine sex. Moreover these sex chromosomes comprise two types, X and Y, which are facts that most readers probably know well. In general, a gene occupies a fixed place on a chromosome, the substance of the gene is DNA (deoxyribonucleic acid), and the gene can be regarded as a fragment of DNA. DNA has a double helix structure as proposed by J. Watson and F. Crick in 1953 (the two intertwined strands each comprise alternating sugars and phosphates linked together, and these two strands are bridged by base-pairs attached to the sugars).

The Flow of Genetic Information What is important from an evolutionary perspective is that DNA is the carrier of genetic information, that genes use four types of DNA bases—adenine (A), thymine (T), guanine (G), and cytosine (C)—as letters, and can be regarded as a coded message written by arranging them in linear order. Specifically, in the DNA double helix, the bases occur as A–T, T–A, C–G, and G–C pairs, but in dealing with genetic information it is sufficient to focus on just one of the two strands of DNA, and therefore in the interest of simplicity I will in the

M. Kimura, *My Thoughts on Biological Evolution*, Evolutionary Studies,
https://doi.org/10.1007/978-981-15-6165-8_4

subsequent discussions refer only to the strand that carries the information (for readers who wish to know more about the structure of DNA and the genetic code, I recommend *DNA and Genetic Information* (Iwanami Pocket Book) by Kin-ichiro Miura.

A gene is many—for example about 1000—DNA bases linked together in linear order, which usually form instructions for making a protein. And when the instructions are implemented, they are copied onto another nucleic acid molecule called messenger RNA. RNA also uses four "letters", and the only difference with DNA is that T (thymine) is replaced by U (uracil). This phenomenon in which the DNA base sequence of a gene is copied on to the base sequence of messenger RNA is called "transcription". The messenger RNA made in this way directs the order in which amino acids are joined together during protein synthesis, and this process is called "translation". Proteins are the most important substances in the body. That is, some proteins are the building blocks of the body, while others function as enzymes that catalyze chemical reactions in the body; it can be said that our existence as biological organisms would be inconceivable without proteins.

Total Length of Human DNA How many DNA base pairs are contained in the human genome (haploid set of chromosomes)? In other words, how many letters (DNA bases) are used to write the genetic commands that are contained in the nucleus of a sperm or egg? The answer as estimated from the amount of DNA in the nucleus is about 3 billion. If the DNA double helical strands forming the 23 chromosomes are connected end to end and stretched in a straight line, its length would be about 1 m, and in a somatic cell nucleus it would be twice this length, i.e., 2 m. And if we join together all of the DNA in the human body (assuming the total number of cells to be 10^{13}), its length would be a surprising 20 billion kilometers, which is approximately 130 times the average distance between the Earth and the Sun (about 150 million kilometers).

Amount of Information Contained in the Nucleus of a Fertilized Egg We may be more interested in the amount of information that is potentially contained in the nucleus of a fertilized egg. In other words, how complex are the commands that can be expressed by the sentences using as letters a total of 6 billion bases transmitted from the two parents.

This can be easily understood if we convert the amount of genetic information into English sentences. Whereas DNA has four types of letters, English sentences use an alphabet of 26 letters, and if a space between words is counted as one letter there are 27 letters. The amount of information is usually expressed with the bit as its unit, which corresponds to a command in which "0 or 1" is chosen. Therefore, because $4 = 2 \times 2$, four letters mean that the location of one base (site) can contain two bits of information. In the case of English sentences the amount of information per letter x can be obtained from $2^x = 27$, which entails that x is about 4.75 (bits). Hence, the amount of information in one letter of an English sentence is the equivalent of about 2.38 letters (bases) of DNA. From this we can see that the amount of information that can be contained in the nucleus of a fertilized egg of the human corresponds to a text comprising about 2.5 billion letters when converted into English sentences.

If we take the Encyclopedia Britannica as an example of a major publication, the 1956 edition comprises a total of 23 volumes, each volume consists of about 1,000 pages, and it can be estimated to contain 200 million letters in all. Thus, if the blueprint for making a human is contained in the nucleus of a fertilized egg, it would when converted to English sentences be of an enormous size corresponding to about 12 sets of the Encyclopedia Britannica combined. From this we can imagine how complex a structure the human is as a biological organism. Moreover, the brain which is constructed according to the genetic commands is a kind of a computer, and the amount of information that is stored over a life time in this brain through learning and experience is also enormous.

Only a Few Percent of DNA Are Genes I wish to mention two caveats here. First, as recent research in molecular biology and particularly in molecular evolution have shown, only a few percent of the DNA in higher mammals apparently function as genes. Most of the remainder is thought to be "junk" without a clear function. Moreover, as explained in Chap. 8, if the neutral theory is correct, in most locations of DNA which base occurs is irrelevant for survival, and there are many places where the text is such that any letter will suffice and the content does not matter (in addition there are many places where the commands inherited from the two parents are identical). Therefore, it may be more appropriate to regard the information content in a usual sense to be the equivalent of one set of the Encyclopedia Britannica (23 volumes). Nevertheless, it is truly an enormous amount of information.

Genes Are the Basis of Life The second point is that the discussion in this section is premised on the thesis that genes are the basis of life. In other words, it is the thesis that the human and other higher organisms grow and become adults according to the genetic commands written in the nucleus of the fertilized egg. This thesis is recently becoming widely accepted through the many advances in molecular genetics. However, it is a thesis that previously was not generally entertained and was strongly opposed especially by researchers of developmental biology.

It was about 40 years ago when I was a student in the Botany Department of a university; I cannot forget that when I tried to persuade a student in the Zoology Department of about the same age as myself of the importance of genetics and said "whether a fertilized egg becomes a mouse or a human is determined by genes", he objected strongly and said that "in lectures on developmental biology we have been taught that such a simplistic idea is wrong". This thesis that genes are the basis of life was originally strongly put forth by H. J. Muller, and when in 1926 he lectured on this at the international conference on botany, the president urged him to change the title of his talk saying that his thesis was too extreme. In those days there were some even among first-class geneticists who claimed that genes were only a concept to explain the results of crossings.

Homeobox As the strongest evidence showing that genes have fundamental control over morphogenesis in animals, I would like to mention a region of DNA called the "homeobox" that was recently discovered by W. J. Gehring.

The word homeobox derives from the term homeotic mutation which was known before from genetic research on fruit flies. This is a mutation in which a part of the body changes into another part, and in particular a mutant gene called *Antennapedia* (*Antp*) which converts an antenna on the head into a foot was the starting point of Gehring's research. The homeobox was discovered as one part of this gene and comprises about 180 DNA bases. This part is found within the base sequences not only of genes in other places of the genome, for example, *bithorax* (which changes abdominal segments into thoracic segments and increases the number of feet) or *fushi tarazu* (which reduces the number of segments of the larva by half), but also in the mating type genes of yeast. Even more surprisingly, genes with a homeobox exist in the human.

The sequences of the approximately 60 amino acids that correspond to the homeobox are virtually identical among the various proteins of which they form a part and over a wide taxonomic range from sea urchin to the human. This is the part of the protein called the homeo domain, and is thought to function as a DNA binding region. That is, it binds to a specific gene and regulates its expression.

It is certain that the group of homeotic genes are the basic genes directing segmental structure in the development of insects. More importantly, the homeobox offers the possibility of providing the first clues in elucidating the mechanism of morphogenesis in higher animals, which has been one of the fundamental mysteries in biology until now. I am keenly interested in future developments in this area of research.

From the standpoint that genes—and more generally speaking the genetic commands contained in the DNA in the nucleus—are the basis of life, biological evolution is none other than change over time of the genetic commands in the species. Its material is genetic variation within the species, and it goes without saying that the environment in which the species is placed is extremely important as the place where natural selection and random genetic drift act to determine the course of change in the genetic structure of the species.

4.2 Nature and Variety of Mutations

Chromosomal Mutations and Polyploidy Mutations are generally speaking sudden changes in the genetic material, and among the various evolutionary factors the most clearly understood due to progress in genetics. According to traditional classification "chromosomal mutations" and "gene mutations" are distinguished, and I will follow this classification for the moment.

First, with regard to chromosomal mutations, these are changes in number and structure. A well-known case of variation in chromosomal number is "polyploidy". This is the case where the number of chromosomes has increased to three sets, four sets, etc., in contrast to the normal individual that has two sets of chromosomes descended from the two parents. In plants tetraploids often arise by the doubling of chromosomes, and moreover there is evidence showing that this plays an important role in the evolution of plants. Especially important are the amphidiploids which arise by doubling of the chromosomes in the first filial generation born of the hybridization of different closely related species. In this case if the genomes (haploid chromosome sets) of the parents are denoted A and B, chromosomes in the somatic cells of the descendant amphidiploids can be written as $AABB$; this is adequately fertile and has the potential to evolve as a species independently of the two parents.

As one example I will mention the research conducted by the late Dr. Tsutomu Haga on the perennial *Trillium hagae*, which is believed to have arisen as a natural hybrid between two *Trillium* species, one with the chromosomal composition K_1K_1 and the other with the chromosomal composition K_2K_2TT (K_1, K_2, and T each represents a genome comprising five chromosomes). Triploid (K_1K_2T) and sexploid ($K_1K_1K_2K_2TT$) forms of *Trillium hagae* are known; the triploid is virtually sterile whereas the sexploid that arose by chromosomal doubling has normal meiosis and bears approximately the same amount of fruit as the parents. The triploid is almost always found, although at low frequency, where the parental species coexist. The sexploid has the property of a stable species, and according to an investigation by Dr. Haga has grown into a fairly large population in one sector of Hokkaido. Other wild plants and cultivated plants are known that are believed to have originated by amphidiploidy.

It is well known that, through the research of the late Dr. Hitoshi Kihara, bread wheat (the sexploid $AABBDD$) was shown to be the amphidiploid of *Triticum diccocum Schuebl* ($AABB$) and the wild plant *Aegilops tauschii* (DD).

We should bear in mind that the evolution by amphidiploid formation is possible only when preceded by speciation (differentiation of genomes). Incidentally, rice which is also a cultivated plant, does not differ in the number of chromosomes from the wild rice believed to be its ancestor, and adaptation in the environment of cultivation apparently played an important role in the evolution of cultivated rice (for details on this, I refer the reader to the article by Professor Keiko Morishima of the National Institute of Genetics cited at the end of the book).

Aneuploidy "Aneuploidy" is the case where several members of a chromosome set increase or decrease in number; in particular the phenomenon in which one extra chromosome is added to the normal complement of chromosomes resulting in three copies of one of the chromosomes is called "trisomy", and many examples are known in plants. Aneuploidy in general perturbs the normal quantitative balance among the various genes of an individual and hence is often disadvantageous for the survival of the individual, and in particular the loss of a chromosome is usually very harmful.

However, in plants the addition of one or two chromosomes often does not adversely affect growth, and in the case of trisomy any one of the chromosomes may be added without compromising growth and seed formation.

By contrast, the situation is quite different in humans; of the haploid set of 23 chromosomes, number 21 which is the smallest of the autosomes (a chromosome other than the sex chromosome) and the sex chromosome are exceptions, but an individual with an extra copy of any of the others cannot survive. Moreover, he or she will be afflicted with a serious defect even when able to survive. The famous example of trisomy in humans is undoubtedly the case of an extra chromosome number 21, which is called "Down's syndrome" (Mongolism) and is usually associated with extreme retardation.

What is the reason for such a difference between plants and humans? A fundamental difference in the structural plan is likely the cause. Plants are essentially factories for the manufacture of starch using sunlight. By contrast, higher animals and in particular the human are precision machines, with a highly efficient computer (brain) which receives information from the outside and controls behavior, and react with much greater sensitivity than plants to small disruptions of the blueprint (i.e., totality of genes) required to make it.

Inversion Next, examples of structural changes of a chromosome are "inversion", "duplication", "deletion", "translocation", etc. An inversion is an abnormality in which part of chromosome has been inverted, and arises through a process in which the chromosome is cut and fused again. The organism in which inversions have been most thoroughly studied is the fruit fly. It has the advantage that not only genetic analysis by breeding is possible, but that inversion can be detected by observation under the microscope of the characteristic bands of the giant salivary gland chromosomes, and intensive and extensive research is being conducted which no other material allows. Chromosomal polymorphism involving inversions were mentioned in Chap. 2 and is a phenomenon often observed in fruit flies and their relatives. There are cases where the same inversion chromosome exists polymorphically in two closely related species, and we can surmise that such an inversion arose in the distant past.

Duplication and Deletion "Duplication" and "deletion" of a chromosome are literally changes in which a part of a chromosome is duplicated or deleted. The size of the change may vary in extent from the case in which nearly a whole chromosome is added or lost to the case discussed later in which the change is minute and essentially indistinguishable from a gene mutation. As an example of a chromosomal abnormality in humans, it is known that in the genetic disease called "cat's cry syndrome" part of the short arm of one of the fifth chromosomes is deleted. It has this name because the patient cries like a cat, but it is also associated with poor growth and mental retardation.

A duplication is a change in which a redundant copy of part of a chromosome is created and, in contrast to a deletion, is generally believed to have in most cases only a small deleterious effect on survival. In particular, a small duplication amounting to one gene has almost no deleterious effect, and is thought in many cases to be neutral under natural selection. What is evolutionally important is the possibility that a new gene will be born from genetic duplication. The importance of gene duplication in evolution is said to have been first proposed in 1936 by C. B. Bridges, based on genetic studies of the fruit fly; but it seems that at about same time, H. J. Muller, based on studies of a mutant called "Bar", formed the view that, when there are two duplicated genes, different mutations may accumulate in them, resulting after many years in genes with different functions. If a gene indispensable for survival exists in single copy in a genome, it cannot readily be changed without compromising survival, but if two copies are made available by duplication, then it is not hard to imagine that new mutations may accumulate and provide the possibility of conducting an "evolutionary experiment". In 1970 Dr. Susumu Ohno in the United States, taking into account many discoveries in molecular biology, published a book on the evolutionary significance of gene duplication, which attracted much attention (Japanese translation *Evolution by Gene Duplication* by Yamagishi and Ryo, Iwanami Publisher, 1977).

Since then, methods have been developed for readily determining the base sequence of DNA, and as the genomic structures of the higher animals and plants were revealed, it became clear that a surprisingly large number of gene families existed in the higher organisms that had arisen by duplication. Rather, genes that exist in single copy may be the exception. As an extreme case, gene families such as ribosomal RNA (a component of ribosomes, which are factories for protein synthesis) and histone (a protein that binds to nuclear DNA in eukaryotic cells), in which several hundred virtually identical genes are repeated, have been discovered. In 1975, L. Hood and colleagues called such a gene family "a multigene family" and pointed out its functional and evolutionary importance. Later, a mathematical theory of the population genetics of the multigene family was developed by Dr. Tomoko Ohta.

Translocation "Translocation" is the phenomenon in which a part of a chromosome breaks off and is joined to another chromosome. In particular, the exchange of chromosomal parts between two non-homologous chromosomes is called "reciprocal translocation". Although I will omit details, it is well known that an individual heterozygous for a reciprocal translocation often cannot allocate a complete set of chromosomes to its gametes during meiosis, and that in plants semi-sterility occurs. Therefore, when two closely related species exist that differ by a reciprocal translocation, the evolution of one of these species must have proceeded by overcoming the obstacle of semi-sterility. Most likely, population size deceased greatly during the formation of the new species, and the reciprocal translocation was fixed by the force of random drift.

Fusion of Centromeres These inversions and translocations cause a change in chromosome morphology. For example, a so called V-shaped chromosome with the centromere (the part of a chromosome where the spindle fibers attach during mitosis) in its middle may change by inversion into an I-shaped chromosome with the centromere at its end, and vice versa.

Another change that I would like to mention is the phenomenon of "centromeric fusion"; here two chromosomes with centromeres at their tips become attached at their centromeres and become one V-shaped chromosome. This is sometimes called "Robertsonian fusion" and is said to frequently occur during evolution. As one example, the chimpanzee, pygmy chimpanzee, gorilla, and orangutan, belonging to the great apes Pongidae, each have 48 chromosomes in their somatic cells, which is two more than 46 in the human. From detailed studies using various differential staining methods, it seems fairly certain that the change from 48 to 46 is the result of two different telocentric chromosomes becoming attached at their centromeres. In general, the region of a chromosome near the centromere comprises genetically inactive material called heterochromatin, and changes of this part have almost no effect on survival. However, if two independent centromeres exist on one chromosome, a tug-of-war ensues during division, which is incompatible with survival, and it is obvious that one of the centromeres must be lost or the two must function together.

That centromeric fusion can occur between homologous chromosomes can be seen by the existence of a strain of fruit flies with an "attached X" in which two X chromosomes have become attached at their centromeres to become V-shaped. This strain is often used in genetic experiments. During the evolutionary process, not only centromeric fusion, but the reverse phenomenon, namely the phenomenon in which one V-shaped chromosome divides into two telocentric chromosomes, will also occur.

It should be noted that such changes do not affect genetic information, but are superficial changes of the kind where, as it were, certain assets are all placed in one safe or are divided and placed in two safes.

Gene Mutation Next, a gene mutation is a change in the internal structure of a gene, which is ordinarily detected by a change of phenotype, and their existence was until now discovered by breeding experiments. However, due to recent progress in molecular genetics, examples are rapidly increasing in which the true nature of a mutation is materially ascertained as a specific change in the internal structure of a gene.

A gene mutation is in molecular terms a substitution, deletion, transposition, or duplication of a DNA base or bases that make up a gene, or an insertion of a base or bases from elsewhere, and can be considered an extremely small chromosomal abnormality. Studies of molecular evolution suggest that the substitution of DNA bases (base substitution) are the most frequently occurring mutations at the molecular level.

Moreover, recent research has shown that a considerable fraction of mutations that are expressed as visible changes of phenotype in higher organisms are caused by a type of "movable genetic element" called an "insertion sequence", which insinuates itself into various genetic loci on a chromosome and changes the phenotypic effects of these genes. I will take this up again later.

Mutations also occur in the nuclei of somatic cells, but what are of significance for evolution are, of course, the mutations that occur in the germ cells and are transmitted to the next generation. In general a gene is highly stable, and mutations occur very rarely. In higher organisms the frequency with which a mutation occurs naturally in one gene (spontaneous mutation rate) is usually about 1/100,000 per generation or less. Nevertheless, in higher animals such as the human, the total number of genes in the nucleus is at least several tens of thousands, so that the probability per generation of a mutation occurring in one of them is non-negligibly large.

4.3 The Nature of Gene Mutation

A Mutation is an "Error" Gene mutations are, along with the self-replicating action of genes, a most fundamental property of biological organisms; almost all traits of all organisms from the human to viruses can change by gene mutation. In molecular terms mutations arise when an abnormality occurs in the double helical DNA chain due to various causes. Then, when the DNA chain replicates, this change is incorporated into the DNA chains of the descendants and is established as a permanent variant (in addition, mutations may occur when chromosomes recombine, i.e., when DNA chains are reconnected, but I will not deal with this here).

In the previous section I compared the gene to a genetic command; using this metaphor the nature of mutation can be well understood if we consider mutations to be various "errors" that occur when this command is transmitted from parent to child. In particular, copying error is important, and our daily experience tells us that mistakes necessarily occur no matter how cautious we are when the text to be copied is enormously long.

The Accuracy of Replication is Amazing Nevertheless, the accuracy of DNA replication of organisms is amazing. Let me demonstrate this by a simple calculation. In the case of the human male, it is estimated that about 30 divisions occur when sperm are made from a primordial germ cell; if we assume that a gene comprises 1000 DNA bases, a gene mutation rate of 1 in 100,000 per generation translates into an error probability of approximately 3×10^{-10} per base per DNA replication. On the other hand, according to a calculation conducted on the basis of the theory of physical chemistry, when a new DNA double helix is made from one DNA double helix, the rate at which the correct base pairing rule of C to G and A to T is not followed, but instead a wrong base is inserted into the new helix, is between 10^{-4} and 10^{-5} per base per replication. In other words, the replication error that is actually observed is 100,000 times more accurate than the value predicted by considerations of physical chemistry.

Repair Mechanisms How is this amazing fidelity achieved? Detailed studies using *Escherichia coli* are showing that this is due to mechanisms for "repairing" errors possessed by biological organisms. In particular, the coordinated action of a group of DNA repair enzymes plays an important role.

In general, DNA is damaged by various internal and external factors. I will leave to a technical book the details of the molecular mechanisms of the induction of mutations and their repair; here, as one example, I will describe the thymine dimer, which is damage to DNA caused by ultraviolet light. When ultraviolet light strikes a part of a DNA chain where two thymines are by chance situated next to each other, the two may adhere forming a dimer (usually expressed as T^T). If this is not remedied, a large gap will form during the next replication in the corresponding position of the newly made DNA chain, and in the worst case the cell may die. However, in a normal cell this dimer is detected by cellular functions. Several bases including this dimer are cut out and removed, and then the correct bases are inserted into this gap using the base sequence of the opposite strand as a template. In this way, the damage to DNA is repaired. This phenomenon is called "excision repair". Xeroderma pigmentosum is a genetic disease in humans in which the patient on exposure to the sun develops red spots on the skin and eventually dies of skin cancer. This disease is caused by the absence in the cells of the patient of one of the enzymes involved in excision repair, and is known to be recessively inherited.

SOS Repair I should like to point out that various mutation repair functions are quite similar to the work of proof reading to correct misprints in the publication process. One other thing I should like to add is that not all DNA repair necessarily corrects errors and sometimes introduces mistakes in the genetic letters, in which case it is rather the cause of mutations.

Well known in connection with repair that introduces such mistakes is the phenomenon called "SOS repair". I will explain this, taking as an example when the cell fails in the excision repair of a thymine dimer; left unattended, the synthesis of the new DNA chain cannot proceed beyond the dimer, and the cell receiving the incomplete chain dies. When the SOS repair function is mobilized to overcome this difficulty, the missing part of the DNA chain is filled with a completely random base sequence, ignoring the rules of base pairing. As a result, mutation occurs. It can be said that this is an adaptive reaction in the sense that to mutate is better than to die. This kind of reaction is likely an important cause of mutations including DNA base substitutions and deletions.

Mutation Lacks Directionality A property of mutations that is particularly important in connection with evolution is that this is a random event at the sub-molecular level, and that, when an organism is exposed to a different environment, a mutation that is adaptive in that environment is not directionally induced. This is an important point that cannot be over-emphasized.

As one example, when a colony of *E. coli* is placed in contact with a specific antibiotic, the bacteria acquire resistance to it after a while. There was a time when two opposing hypotheses were proposed to explain this phenomenon. The first is the claim that contact with an antibiotic induces a resistant variant as an adaptive physiological response of the bacteria. The second is the claim that, among the randomly occurring mutations unrelated to antibiotic use, there is by chance a mutation that confers resistance on the bacteria, and that it is the consequence of only the bacteria carrying this mutation surviving and reproducing in the presence of the antibiotic (in other words, of selection).

For some time directly after the war, when I was a college student, these two hypotheses were also vigorously argued in Japan, and the first hypothesis in particular was strongly supported by biologists who were skeptical of Mendelian genetics and orthodox evolutionary genetics based on it, including followers of Lysenkoism. However, several decisive experiments were subsequently conducted, and it was proven that this first hypothesis was wrong and the second hypothesis was correct.

A Decisive Experiment Famous among these is the experiment by J. Lederberg and colleagues using the replica plating method. The main points are as follows. First, the agar medium in a Petri dish is seeded with bacteria, and colonies are allowed to form. Next, a cotton velvet cloth is affixed to the flat end of a cylinder, and each colony is made to adhere to this flat end by pressing it against the surface of the medium exactly as one would press a stamp on a pad of ink. Then, this is transferred to a second agar medium as one would press the stamp on paper, and the colonies are allowed to grow on this second medium as well. What is important is that bacteria of identical origin (i.e., descended from the same colony) exist on these two mediums at corresponding positions. After this, an antibiotic is added to one of the mediums, and if a surviving colony eventually emerges, by taking the colony at the corresponding position on the medium in the other Petri dish and culturing it, bacteria with resistance to the antibiotic ("resistant bacteria") would be found without exposure to the antibiotic. This experiment showed that mutations conferring resistance arose at a constant rate irrespective of the existence of an antibiotic. Such a mutant is less competitive than a nonresistant individual in a normal environment and cannot multiply, but in the presence of an antibiotic nonresistant bacteria die, and the former rapidly multiplies to replace the latter. Bacteria acquire resistance for this reason.

Similarly, Professor J. F. Crow using the fruit fly succeeded in obtaining by sib-selection a strain with resistance to an insecticide without exposure to that insecticide. Here sib-selection is a selection method in which the siblings of the individuals in question are used and is the method that has often been used in animal breeding, because when strains of beef or chicken with tasty meat are to be selected and reared, they cannot be obtained from the individuals that are killed to taste the meat. Experiments on the acquisition of insecticide resistance are important in showing that resistance genes are arising by random mutation in a population of flies not exposed to the insecticide.

As an example similar to the acquisition of resistance to an antibiotic in bacteria, we can mention the phenomenon, studied in Japan, of the acquisition of resistance to copper sulfate in yeast. This was also thought at first not to be due to mutation, but rather to be an adaptive physiological response of yeast to copper sulfate, but it was proved by the genetic studies of Dr. Takuji Seno that this phenomenon was also based on the appearance of a (dominant) mutation conferring resistance, followed by selection.

4.4 Phenotypic Effects of Gene Mutations

Phenotypic Effects Are Diverse As already mentioned, gene mutation is one of the most fundamental properties of biological organisms, and almost all traits of all organisms may change by gene mutation. In this respect, the fruit fly is the most thoroughly investigated higher organism, followed by the human.

An enormous number of mutations have been described in the fruit fly, from morphological characters such as eye color, body color, number and form of bristles, wing shape, size and shape of the eye, to purely physiological characters such as the response to light and gravitation. Moreover, the changes are truly diverse in their extent, ranging from lethality and mutations with striking effects such as eyeless and the presence of four wings instead of the normative two, to those that lower viability very slightly and no abnormality is superficially observable. Furthermore, although not an example in the fruit fly, a mutation advantageous for survival over the preexisting wild type, such as the melanic from in the industrial melanism of moths, can also arise as discussed in Chap. 2.

In addition, most of the DNA base substitutions (or the resulting amino acid changes), their detection being made possible by molecular genetical methods, usually cause no change to the phenotype and can be regarded as having no effects on survival or reproduction, in other words, as being mutations that are neutral under natural selection. If the neutral theory is correct, this type of mutation should be occurring quite commonly and widely among organisms.

"Mutations Are Deformities" Is a Popular Belief Here I would like to emphasize that the popularly held belief that "mutations are striking deformities" is mistaken. Even now there are some even among biologists who dogmatically aver that mutations cannot serve as material for evolution because they are all deleterious, and the fanatical among them go as far as to criticize and to attack orthodox evolutionary genetics as the "evolutionary theory of deformities". These people are probably under the spell of the popular belief that "mutations are deformities".

One reason why such a belief became prevalent may be that genetics textbooks in the past had as their chief aim the explication of the modes of inheritance (transmission mechanisms of genes), and to achieve this goal, to always cite as markers mutant genes with phenotypic effects that were strikingly different from the normal genes, and to pay little attention to the overall properties of mutant genes. In general, a mutation is more likely to be discovered and described when it causes a conspicuous difference from the normal wild type gene, and conversely a mutation causing a minor difference is more likely to be overlooked as a change due simply to the environment. However, this type of mutation of minor effect is not essentially different from the former, and in particular in no way inferior in its stability and permanence. Indeed, these mutations of minor effect are the important material for evolution.

Quantitative Characters In this connection I would like to mention the so-called quantitative characters, such as body size and the proportions of a leaf. When two closely related species are crossed, it is known that in the first filial generation (F_1) the range of individual variation for such a character is approximately the same as among the parents, but that in the second filial (F_2) and subsequent generations it suddenly increases.

The reason for this is that the parents have differentiated by accumulating many mutually different mutations of minor effect, which segregate regularly in the F_2 and subsequent generations. In other words, if the alleles coming from the parents are denoted *A* and *a*, all F_1 individuals are *Aa* whereas in the F_2, *AA*, *Aa*, and *aa* segregate in the ratio 1:2:1; and because this occurs at each of many genetic loci, the number of genotypes formed by the combination of different loci will be enormous. The phenotypic differences due to the individual allelic pairs mix with variation originating from environmental conditions to generate the continuous individual variation in the population. Because such mutations are evolutionarily important, I would like to take them up in a little more detail.

Many Genes Are Involved Generally speaking, if ontogenesis from fertilized egg to adult proceeds in accordance with the genetic commands in the nucleus, the action of the set of genes on the expression of characters is as it were an intricate network. Therefore, it is a reasonable assumption that in many cases many genes are involved in one character, and that one gene often affects directly or indirectly the expression of two or more characters. And it is natural for a change in what may be thought to be the "normal" range to occur by mutation. In fact, many genes are involved in what in genetics is called a "quantitative character", such as height, weight, intelligence quotient in the human, quantity of milk produced by a dairy cow, yield of rice; and moreover the difference in the effects of the alleles at each genetic locus is small and in fact much smaller than the range of individual variation due to the effect of environment.

One feature of a quantitative character is that the environment also has a large effect on the phenotype. As such, the Mendel-Morgan genetic analysis which examines the conventional segregation ratios can be applied with difficulty. As a result, there were previously some people who claimed that gene theory was not applicable to a quantitative character showing continuous variation. However, subsequently with progress in statistical genetics, a scientific method for its analysis was developed, and currently it has become one of the evidences that prove the truth of gene theory. The behavior of the alleles involved in a quantitative character of course obey Mendel's Law of Segregation.

These genes for a quantitative character were at first called "multiple factors"; later the term "polygene" proposed by K. Mather came to be generally used. However, recently, the expression "quantitative trait loci", often abbreviated QTL, is widely used, focusing on the genetic loci that are involved. Quantitative characters include many that are selected in animal and plant breeding and are of economic importance, such as milk yield in cows, body weight in pigs, number of eggs laid by a chicken, and yield of corn; and the genetics of quantitative characters is an important area of basic research in breeding studies.

Heritability I will not treat in detail the subject of statistical genetics dealing with quantitative characters, but I would like to mention one important concept. This is what is called "heritability", a quantity that expresses the degree to which a given quantitative character in a specific population is genetically determined. In a simple case, this can be obtained by estimating the correlation coefficient for this character between parent and child and then doubling this value (rigorously speaking, heritability is defined as the fraction of the total phenotypic variance for the character that is due to the additive effects of the genes. Here, additive effects mean the statistically obtained average effects of the genes after removing the non-additive effects of the genes such as dominance relations and epistasis).

As examples of heritability, values of 0.2–0.4 for milk yield in cows and about 0.2 for the number of eggs laid by a chicken have been reported. For human height a value greater than 0.5 has been obtained, which is much higher than in these two examples (Dr. Toshiyuki Furusho has conducted detailed studies on the heritability of height, and his paper cited at the end of this book will provide useful information). In general, the higher the heritability the easier it is to change the character by selection in the desired direction, and hence heritability has a very important meaning in breeding.

Movable Genetic Elements Finally, I would like to mention the "transposable element", in other words the intra-locus insertion of a movable genetic element, whose importance as a cause of mutation at the phenotypic level is recently becoming clear. As briefly noted in Sect. 4.2 of this chapter, this is in short the phenomenon in which an insertion sequence or transposon—i.e., an insertion sequence containing a gene—moves around the chromosome and inserts itself in various genetic loci, as a result of which the phenotypic effect of a gene changes and is expressed as a mutation.

Among the higher organisms this phenomenon has been studied most in the fruit fly, and in its nucleus an average of about 50 transposable elements called "copia" exist per genome. In addition, almost 30 kinds of similar transposable elements are already known, which are collectively called "copia-like elements"; they enter various genetic loci on the chromosome, and not only do they cause "visible mutations" such as changes of eye color, but also induce mutations lowering the viability of the fly or mutations affecting a quantitative character like the number of bristles. It is said that approximately half of the spontaneously occurring mutations in the fruit fly are due to the insertion of a copia-like element. Nevertheless, the number of transposable elements per genome is much smaller than the total number of genetic loci, and moreover the rate of transposition is low, so it remains true that mutations due to this cause are generally speaking a rare phenomenon.

Whether or not the same can be said about mammals like the human and the mouse must await future research.

On Natural Selection and Adaptation 5

5.1 Darwin on Natural Selection

The Origin of Species As the starting point for our discussion of what natural selection is, it is convenient to consider Darwin's famous book *The Origin of Species*. The full title of this book including the subtitle is *On the Origin of Species by Means of Natural Selection, or the Preservation of Favoured Races in the Struggle for Life*. We can gather from this that the theory of natural selection forms the basis of Darwin's theory. This book was published when Darwin was 50 years old; stimulated by A. R. Wallace's paper on natural selection, the results of his researches, tirelessly continued for 22 years after returning from the circumnavigation of the globe on the Beagle, were made public in the form of a compendium.

In what follows, I would like to explain his theory of natural selection and the ideas that formed its basis as set out in the first edition of *The Origin of Species.*

In the Introduction to this book, Darwin states that in order to show that all species were not separately created but are derived by evolution from a common ancestor, it is necessary to explain how all species came to be surprisingly adapted to their environments. In other words, his position is that it is meaningless simply to show that species change, without simultaneously explaining the adaptations widely existing in the biological world (this point is also clearly mentioned in his autobiography).

Variation in Domesticated Animals and Plants Darwin sought the clue to solving this problem in the studies on the variation of domesticated animals and plants. In Chap. 1 which deals with this, he first of all draws the reader's attention to the existence of much greater variation among the individuals belonging to the same variety of a domesticated animal or cultivated plant than under natural conditions. In order to investigate in detail how the differences between different varieties are formed, Darwin chose the domestic pigeon as the material for his research, and himself joined two pigeon lovers' clubs in London. Domestic pigeons include the carrier pigeon, the tumbler, and the fantail, which are varieties that are surprisingly

M. Kimura, *My Thoughts on Biological Evolution*, Evolutionary Studies,
https://doi.org/10.1007/978-981-15-6165-8_5

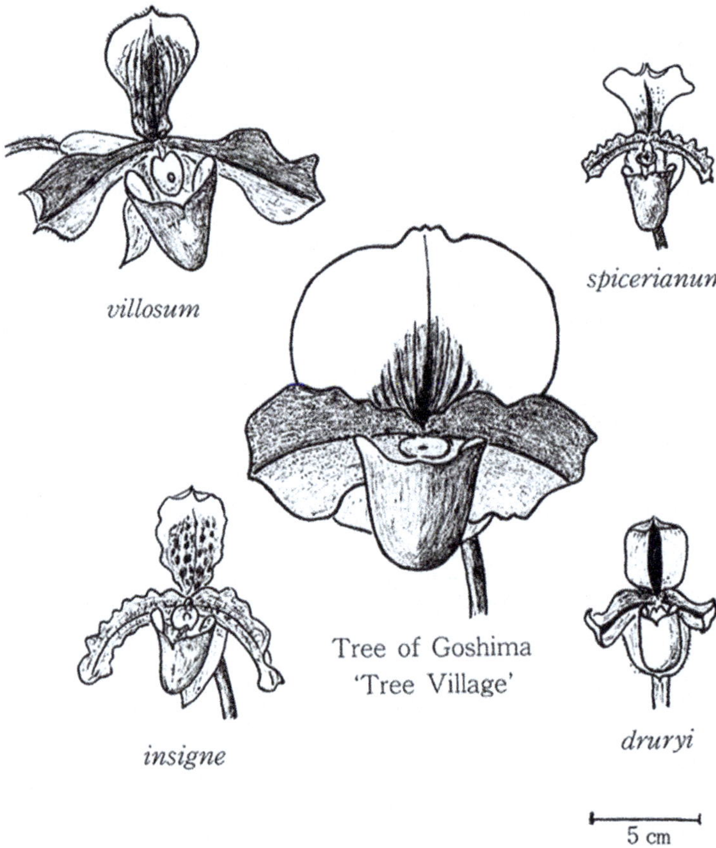

Fig. 5.1 Shows the remarkable modifications in the size and form of *Paphiopedilum*, a genus of orchids, achieved by several generations of crossing and selection. The flower in the center is 'Tree Village', a variety of a modern horticultural species, *P. tree* of Goshima (=Zushi × Borburn, registered by Kimura 1976). In the four corners are shown the flowers of the four ancestral (wild) species several generations removed, *P. villosum* (above left), *P. insigne* (below left), *P. druryi* (below right), and *P. spicerianum* (above right)

different in appearance and habits. Darwin thought that if these were feral birds, ornithologists would undoubtedly classify them not only as different species but as different genera. However, as his researches progressed he was compelled to accept as correct the opinion of amateur naturalists that these varieties all had the feral species *Columba livia* as their ancestor and were descended from it (Fig. 5.1).

How then did these major differences arise? It is nearly impossible to attribute this to the direct effects of the external environment or to habit. The key to the solution can be sought in the process whereby from among the fortuitously occurring variants, those that conform to the goals of humans are successively selected and accumulate. That the force of "selection" is not imaginary can be seen from the fact that famous breeders have during their lifetimes greatly improved cattle and sheep, and that various flowers have been rendered unrecognizably splendid by horticulturalists in 20–30 years. Of course, it is only recently that selection has been carried out with a specific goal, but unintentional selection was likely carried out even in the past by uncivilized people. If we take the domestic pigeon as an example, selection could not have been pursued from the beginning with the present day fantails as the goal. Rather, after an individual with a slightly abnormal development of the tail was accidentally discovered, it occurred to people to select in this direction.

"Variation Under Nature" Can the principle of selection that gave rise to such diverse varieties of domesticated animals and cultivated plants then be applied to biological organisms in nature? In Chap. 2 entitled "Variation under Nature", examples of individual variation on which natural selection may act are variously described, and it is pointed out that variation exists in characters that taxonomists consider to be important.

In relation to this, he refers to the existence of genera comprising abnormally variable species, referred to as "protean" or "polymorphic" at that time. Next, the fact that some varieties are treated as such because intermediate forms exist, although in many points they possess the features of a species, is important. Moreover, taking into account the cases where intermediate forms are absent but are treated as varieties because the differences are small, it cannot be believed that there is an essential difference between a species and a variety. In addition, in large genera that contain more than the average number of species, there is a tendency for each species to on average contain many varieties, which can be readily understood if we consider that the species were previously varieties and arose from them.

"Struggle for Existence" Chapter 3 of *The Origin of Species* has the title "Struggle for Existence"; Darwin argues that the "struggle for life" is ubiquitous in nature and that this is the cause of "natural selection" discussed in Chap. 4. In Japanese, "struggle for existence" is more often translated as "competition for existence", and in this book they will be used synonymously.

Darwin emphasizes that he uses the term struggle for existence strictly in a broad "metaphorical sense". Therefore, this does not necessarily refer only to two carnivores fighting over one prey animal when hungry. This includes the dependencies between biological organisms and success or failure in leaving offspring (Darwin says this is more important).

The cause of the struggle for existence is the tendency possessed by all biological organisms to increase their numbers geometrically. This is Malthus' theory of population extended and applied to the whole biological world. As is well known, according to Malthus, because population increases geometrically whereas the production of food increases only arithmetically, difficulties will eventually occur as a result of a shortage of food, and population growth will cease.

Organisms have this tendency to increase geometrically, and Darwin showed by a numerical example that even a slowly reproducing species like the elephant has the potential to reach a surprisingly large number in a relatively short time if all offspring survive. In addition, he gives a calculation due to the botanist Linnaeus that, if two seeds from an annual plant develop per generation, the original individual would 20 years later become a million individuals.

If, in each species, it is impossible for the number of individuals to increase indefinitely, then even if many offspring are produced every year most of them are fated to not complete growth and to die. Taking as an example an annual plant that produces 1000 seeds per year per individual, only one on average among them completes growth. As a result, competition for existence necessarily occurs. It is clear that all species are exposed to severe environments from the fact that there are explosive outbreaks of particular animals in years when the environment improves. Of course, the interactions among biological organisms in nature are immeasurably complex, and it is impossible in individual situations to see the whole picture; its complexity can be inferred from the sequential dependencies where the number of bumble bees that pollinate red clover in England is affected by the wild mice that eat their larvae, and the number of wild mice is influenced by the number of cats that live in the area. Competition for existence is most intense between individuals of the same species which eat the same food, live in the same places, and are exposed to common enemies. Moreover, competition is in many cases just as intense between varieties belonging to the same species, as can be seen from the fact that, when different varieties of wheat or snow peas are grown together, the contest is decided in less than a few years.

"Natural Selection" How then does the struggle for existence explained in this way in Chap. 3 affect individual variation? Does the principle of selection really apply to organisms in nature in the same way as artificial selection? In Chap. 4 entitled "Natural Selection" Darwin discusses this problem in detail and endeavors to prove that the principle of selection works very effectively in nature as well. This chapter can be said to form the core of *The Origin of Species*.

Darwin believed that, under the intense struggle for existence described above, if variation advantageous for a biological individual no matter how small occurred, then the individual showing this variation would undoubtedly have a higher viability and produce many descendants. Moreover, such an advantageous property would in general be transmitted to descendants. Darwin says "This preservation of favorable variations and the rejection of injurious variations, I call Natural Selection" (Actually, already in Chap. 3 in connection with the struggle for existence he writes "I have called this principle, by which each slight variation, if useful, is preserved by the term of Natural Selection".

There is no doubt that individual variation exists under natural conditions. As we see variation useful for humans emerging under conditions of cultivation and breeding, it would be extraordinary if variation advantageous to the individual did not appear under natural conditions. And in the process of the struggle for existence, variation that is even slightly advantageous for survival would be preserved and transmitted to descendants by the "strong law of inheritance". Although as in the case of artificial selection the process of change cannot be witnessed, by the continuous action of natural selection over long geological time spans, biological organisms are over time improved to adapt to the environment. Darwin recognized that the action of selection would also be applied to a male's ability to acquire a female, especially in the higher animals, and named this "sexual selection" to distinguish it from the ordinary action of selection. In those cases where male and female of the same species generally share the same environment and habits, but there are large differences between them in body color, ornaments, or body structure, he believed that this difference arose by sexual selection.

Natural selection also has the effect of causing the divergence of species. This is because more individuals can be supported in the same area that differ in structure, habits, etc. than a collection of uniform individuals. Therefore, a species that gives rise to many varieties is more likely to win in competition with other species. Eventually these varieties develop into different species, and at the same time species that lose go extinct. By this principle, the fact that family relations exist between all organisms can be explained. The fact familiar to us that similar varieties together form a species, similar species together form a genus, and that genera can be grouped together into families is, when we think about it, a truly amazing phenomenon.

The text of *The Origin of Species* ends at Chap. 13 (Chap. 14 is "Recapitulation and Conclusion"), and many problems related to natural selection are discussed with extreme care. For example, Darwin considers the following.

What are the circumstances most favorable for the formation of a new species by natural selection? Isolation is certainly effective, but a large range is more important. In particular when continents rise and sink, intense competition occurs in the uninterrupted range when it has risen; and when it sinks and divides into many islands, differentiation into regional species adapted to each of the islands occurs; and when such a process of differentiation and competition is repeated, new species are effectively formed. This part is interesting when compared to Wright's shifting balance theory mentioned in Chap. 2 of this book; namely, the claim that subdivision of a species into incompletely isolated subpopulations allows random differentiation of, and competition between, subpopulations to work effectively together and is most favorable for the evolution of the species as a whole.

Darwin's Deeply Impressive Attitude What made a particularly deep impression on me is that Darwin includes a Chap. 6 entitled "Difficulties on Theory", in which he deals with observations and objections that constitute difficulties for his theory and discusses them head on. Generally speaking, even in the natural sciences which should be objective, it is usual for most authors when they propose a new theory to write only what is convenient for their theories, and subsequently when reiterating their theories to not heed criticisms or to ignore the difficulties. By contrast, it must be said that an attitude of revealing things that are inimical to one's theory and examining them in earnest, as Darwin does, is surprising. It is quite exactly the opposite attitude to "dogmatism". In my limited experience, this way of doing things is seldom seen in Japan or Germany. It may be that the spirit of the philosopher Francis Bacon—who proposed at around the beginning of the seventeenth century that preconception should be discarded and nature should be correctly understood based on experience (experiment and observation) and using induction—or of a predecessor is strongly entrenched in England.

In connection with this, I would like to take up two claims that Darwin made in his autobiography that I was especially impressed by. According to him he had the habit of wanting to set up a hypothesis right away for any problem. However, he continuously endeavored to maintain an open mind, so that no matter how attached he was to a hypothesis, he would be able to discard it immediately once it was found to be contrary to fact. He says that he also followed the golden rule of immediately making a note without fail when he encountered a published fact or observation or idea that contradicted a general result obtained by himself. Darwin's scholarly attitude is apparent in the careful discussions done from a wide perspective that characterize *The Origin of Species*.

Four Difficulties Returning to the main issue, in Chap. 6 he classifies the difficulties of his theory into the following four. First, if a species has evolved gradually from another species, why are not intermediate forms found ubiquitously? In fact, species can be distinguished relatively clearly. Second, is it possible that an animal with a structure and habits like the bat, for example, can be derived from another animal with entirely different habits? Can we really imagine that natural selection would on the one hand create an organ of low importance like the tail of giraffe that is useful only as a fly-swatter, whereas on the other hand it would produce a superbly perfected organ like the eye? Third, can instinct be acquired and modified by natural selection? What can be said about the remarkable instinct that induces the bee to make a splendid nest conforming to mathematical laws? Fourth, why are crosses between different species unproductive or result in sterile descendants, whereas crosses between different varieties do not adversely affect fertility?

Of these four difficult questions, the first two are dealt with in Chap. 6, and the remaining two questions are discussed in Chaps. 7 and 8.

Evolution of the Wing of a Bat I cannot here consider all of these various questions, but as one example I would like to present Darwin's explanation from Chap. 6 of the evolution of the wing of a bat. It is impossible to directly answer the question of how the bat actually evolved—through what kind of intermediate stages—from an insectivorous quadruped, but the question itself is not at all fatal to the theory of natural selection. On the contrary, this difficulty is much reduced if we consider the following example.

If we look at the animals of the *Sciuridae*, they exhibit many minutely different stages, from those with a slightly flat tail to flying squirrels in which the posterior part of the body has widened into a patagium at the side of the body. In particular the patagium in the flying squirrel extends between the four limbs and as far back as the base of the tail, and this acts as parachute allowing the animal to glide surprisingly long distances from tree to tree.

The intermediate structures in these stages can all be considered to have been useful to the species possessing them. For example, in one stage it may have been convenient for rapidly collecting food, and in another it may have been useful to escape from enemies or to reduce the dangerous consequences of an accidental fall from the trees. Under the circumstances in which the environment changes due to transitions of climate and flora or due to the invasion of enemy organisms, it is not at all unreasonable that individuals with even a slightly enlarged patagium would have had a tendency to leave more descendants and that the flying squirrel would eventually evolve due to the cumulative effects of natural selection. Moreover, if we look at *Galeopithecus* (a member of Dermoptera), which was mistakenly classified as a relative of the bat, the patagium at the side of the body is very wide, extending from the corner of the jaw to the tail and covering the long and slender four limbs including the fingers. And this patagium has extensor muscles. Hence, if this kind of structural plan progresses further, it is not at all mysterious that an organism should emerge like the bat with developed wings. Perhaps, we will also find in the ancestor of the bat the vestige of an organ that shows that it began from a structural plan for gliding through the air rather than for flying.

Instinct The question of "instinct" is taken up in detail in Chap. 7; this chapter is full of interesting topics such as the cuckoo's habit of laying its eggs in the nests of other species, species of ants that use other species of ants as slaves, and the instinct of honeybees that make remarkable nests. The establishment of these instincts, if we look only at the completed stages, would appear inexplicable by natural selection theory; but when well-researched there is in each case a closely related species at a highly incomplete developmental stage, intermediate forms are found, and it becomes clear that the difficulty is not a fundamental one.

Here Darwin notes that the evolution of the neuter insects (for example, worker bees) in the social insects cannot be explained by Lamarckism. This is because neuter insects lack reproductive ability and cannot transmit the use and disuse of their desires and organs to their descendants. On the other hand, according to natural selection theory, this can be explained by invoking selection on the female and male individuals that produce the neuter insects. Darwin states that it is a difficult question how traits related to sterility in the social insects evolved gradually by natural selection. Nevertheless, it is deeply impressive how he explains that this apparently insuperable difficulty can be reduced or overcome if we consider that "selection may be applied to the family, as well as to the individual".

Altruistic Behavior Subsequently, after about 100 years, it was demonstrated quantitatively by W. D. Hamilton, using the methods of population genetics, that the genetic disposition to sacrifice oneself for kin would under certain conditions be selectively advantageous. In sociobiology, which was recently proposed by E. O. Wilson and currently attracts the attention of many biologists, the evolution of traits that superficially contradict natural selection theory, in which an individual sacrifices itself to contribute to the survival and reproduction of another individual, in other words of "altruistic behavior ", has become one of the central subjects. Here the term "kin selection" coined by Maynard Smith is often used to denote selection on a kin group rather than selection on an individual. Given these developments, Darwin's foresight is amazing. It should be noted here that efforts made by a parent for its offspring are usually not regarded as altruistic behavior. This is because protection of its offspring is naturally adaptive behavior that raises the fitness of the parent.

For the reader interested in the question of the evolution of the altruistic behavior, I recommend *Biology of Altruistic Behavior* (Kaimeisha 1983) by Kenichi Aoki. This is a short but good book which explains the central problems of sociobiology, including recent research, from the standpoint of population genetics.

"Hybridism" Chapter 8 of *Origin of Species* is entitled "hybridism", and deals with the phenomenon in which descendants are usually not produced when organisms that are regarded as of clearly different species are crossed, or the hybrids when produced are sterile. He argues that this is not a property gradually acquired by the direct action of natural selection, but rather is a property brought about secondarily by chance and incidental on changes of other traits by natural selection, and gives various evidences in support of this claim. This question is deeply connected to speciation. Speciation or species divergence is an important topic in evolutionary theory, and various theories are even now being proposed.

5.2 Modern Developments in the Theory of Natural Selection

Development of the Theory of Natural Selection Historically speaking, the theory of natural selection was first proposed in 1858 by Darwin and Wallace at the Linnaean Society in England, but it seems their papers did not receive much attention in academic circles. It was the publication in the following year of *The Origin of Species* by Darwin that had a global impact, the content of which was described in the previous section.

Subsequent progress and development of the theory of natural selection proceeded mainly in connection with knowledge of genetics. In particular, the quantitative treatment of natural selection made remarkable progress as one of the central subjects of population genetics from the 1920s to the 1930s, as was mentioned in Chap. 2 in connection with the contributions of R. A. Fisher, J. B. S. Haldane, and S. Wright. Needless to say, many advances have been made since then. It is not the purpose of this book to delve deeply into the contents of these researches, so I will here limit myself to briefly explaining the basic concepts.

Fitness In current terminology, natural selection is said to operate when individuals with different genotypes are present in the population, and differences exist between them in viability or fertility (fitness in general). Here, the "fitness" of an individual is a quantity that expresses how many offspring the individual can leave in the next generation (contribution to the making of the next generation). In population genetics it is usual to define the fitness of a genotype as the average number of children per individual for a group of individuals that have a certain genotype. Of course, the number of children refers not simply to the total number that are born, but to the number that reach maturity. Also, in measuring fitness, it is important to count the number of offspring at the same developmental stage as when the parents were counted.

As such, the definition of fitness is relatively simple, but it is not easy to actually estimate this quantity. More specifically, a different measure must be used in the case where generations are discrete as in annual plants, and in the case where individuals of different generations coexist in the population and the generation structure is continuous as in humans. In the latter case, it is appropriate to use the Malthusian parameter proposed by Fisher, i.e., the rate of geometric (exponential) increase or decrease of the number of individuals (usually expressed by the symbol m). However, to obtain this a complicated calculation is necessary involving functions for age specific viabilities and fertilities. In humans it is usually possible to estimate fitness roughly by the number of grown-up daughters per mother.

Next, for a population with discrete generational structure, it suffices as already mentioned to express fitness by the average number of children; in population genetics this is called "selective value" following Wright and is often expressed by the symbol W. As shown in the following footnote, the two cases can be unified using a small approximation.[1]

Increase or Decrease of Alleles in a Population Moreover, in order to compute the rate and to follow the process of the increase or decrease of alleles in a population using the selective values of the genotypes, the absolute values of selective values are in most cases unnecessary and it suffices to know their relative values. Therefore, it is usual to set the selective value of one of the genotypes to 1.

As one example, I will show the calculations in the simplest case of "genic selection", i.e., the case in which selection can be regarded as acting directly on the genes. Let us assume two alleles A and A' in a large population, where A is the preexisting wild type (normal) gene, and on the other hand A' is an allele that arose from it by mutation and is advantageous under natural selection (this represents the case in which genotypes A and A' exist in a haploid population and there is competition between them). As for the fitnesses, when the selective value of A is 1, that of A' is denoted $1 + s$ (here, s expresses the advantage of A' and is called the selection coefficient). In the usual diploid population, the corresponding case is when the selective values of the three genotypes AA, AA', $A'A'$ can be expressed as 1, $1 + s$, $1 + 2s$, respectively. In either case, it is assumed that s is a very small positive number such that s^2 can be ignored compared to s, and the calculations are done accordingly. Then, it is possible to derive the following simple formula showing how the ratio of the frequency of A' and the frequency of its allele A in the population changes as the generations pass.

[1]Although involving an approximation, it is convenient in the case when the generations are continuous to take the average reproductive age as the length of one generation and to relate the Malthusian parameter, m, to the selective value, W, by the formula $e^m = W$, or taking the natural logarithm of both sides by $m = \ln W$.

$$\ln (p_t/q_t) = \ln (p_0/q_0) + st \tag{5.1}$$

Here, ln denotes the natural logarithm. And p_0, q_0 are the fraction (i.e., frequency) of A' and the fraction of A, respectively, in the population. Moreover, the relation $q_0 = 1 - p_0$ holds. p_t, q_t are the corresponding quantities in generation t.[2]

An Advantageous Mutation Spreads Rapidly Formula (5.1) is based on the assumptions that population is extremely large (theoretically infinite), that random fluctuations of p can be ignored, and that s is constant over generations; it is however convenient for computing under these conditions the rate at which a selectively advantageous mutant gene will spread in the population. For example, when A' has an advantage in natural selection of 0.1% over A ($s = 0.001$), the number of generations required for the frequency of A' (p) to increase from 0.1% ($p_0 = 0.001$) to 99.9% ($p_t = 0.999$) is 13,813.5. If one generation is one year, this entails that about 14,000 years are necessary. If the advantage is 1%, which is 10 times the above, the required generations are approximately 1400 years. These times are long compared to our lifetimes, but are very short on the geological time scale.

From this it can be said that if a mutant gene clearly advantageous for the survival or reproduction of an individual appears in the population (assuming that it escapes random loss), it will replace the preexisting wild type gene in a relatively short time

[2]Formula (5.1) is not difficult to derive. I will show it for the interested reader. Set the fraction (i.e., frequency) of A' in the population to p and the frequency of its allele to q ($q = 1 - p$). A' increases in frequency at a constant rate relative to A, and if we note that $\ln(1 + s) \approx s$, then the ratio of the frequencies p/q increases exponentially at rate s, and, if we write p/q in generation t as r_t, then we have $r_t = r_0 e^{st}$. Here r_0 is the initial (in generation 0) value of the ratio p/q. Hence, using subscript t to specify the various quantities in generation t, and setting the natural logarithm of r_t ($\ln r_t$) to z_t, we have

$$z_t = z_0 + st \tag{5a}$$

and we have re-derived formula (5.1). Here,

$$z_t = \ln (p_t/q_t) = \ln \{p_t/(1 - p_t)\}$$

This can also be obtained in the following way. First, directly computing the rate of change of p per generation and approximating it by a differential equation, we get

$$\frac{dp_t}{dt} = sp_t(1 - p_t) \tag{5b}$$

Of course, this formula can also be simply derived from the fact that ratio r_t increases exponentially at the constant rate s and therefore that $\frac{dz_t}{dt} = s$ holds. Next, if we solve this equation with the initial condition that p_t equals p_0 when $t = 0$, then the formula obtained for p_t will be the same in substance as (5.1) or (5a). I refer the reader who wishes to learn about this kind of problem in a little more detail to Chap. 2 of Iwanami Lectures Current Biological Science 6 *Fundamentals of Human Genetics* (edited by Motoo Kimura, 1975).

in the evolutionary process, and will newly acquire the status of the "normal" wild type gene. If such a process is repeated in a given environment, at each genetic locus the mutations that can occur are so to speak tested, one after another, and the most suitable one will be what remains. And the store of mutant genes that have never appeared and are more advantageous for the survival or reproduction of an individual than any that have appeared until now will gradually be exhausted. What we call wild type genes have in all biological species been established by such a process of selection over the past millions of years, tens of millions of years, or even longer times.

From this standpoint, the observed fact that the so-called "mutations" discovered in the laboratory or in natural populations are mostly more or less harmful for the organism is not at all surprising. In other words, such facts strongly support rather than contradict the view that mutations are the material of adaptive evolution (it is important not to confound process and result). Neither does this contradict the view of the neutral theory that wild type genes in the same species are not necessarily uniform in molecular terms but comprise variant alleles that are equivalent (neutral) in fitness.

Fundamental Theorem of Natural Selection R. A. Fisher, at the beginning of his famous book *Genetical Theory of Natural Selection* (1930), states that natural selection and evolution are not synonymous and emphasizes the significance of separating the question of natural selection from evolution and of studying it as an independent subject. I believe the same thing applies to population genetics. In other words, I feel it would be unfortunate if population genetics were to be viewed simply as one part of evolutionary research and its value were to be judged only in relation to evolutionary theory. Of course, to argue about evolutionary mechanisms ignoring the theory of natural selection and knowledge of population genetics is a dangerous way to conduct science, similarly to arguing about the structure and evolution of the universe while ignoring physics.

Fisher's book was at that time an epoch making one and contained many important and interesting results. One of them is what he called the "fundamental theorem of natural selection", and in his words "[t]he rate of increase in fitness of any organism at any time is equal to its genetic variance in fitness at that time". That is

$$\text{rate of increase in fitness} = \text{genetic variance} \tag{5.2}$$

So long as Fisher's fundamental theorem, i.e., formula (5.2) holds, the mean fitness of the population keeps on increasing unless the right hand side, which is a variance (sum of the squares of deviations from the mean) and cannot be negative, is zero. This is in accord with the view that biological evolution in the past was mostly adaptive, entailing the improvement of functions. We can see that Fisher attached great importance to this theorem from the fact that he regarded it as a basic law of biology comparable to the second law of thermodynamics (law of increase of entropy). However, judged calmly half a century later, I do not think it so important. Moreover, I feel it has not contributed to actual evolutionary studies.[3]

Equation for Artificial Selection On the other hand, an equation analogous to (5.2) used to predict the result of artificial selection in animal breeding,

$$\Delta G = h^2 \Delta P \tag{5.3}$$

would seem to have much more utility value. The left hand side of this formula, ΔG, is the increase in one generation of a quantitative character subject to improvement by artificial selection (for example yield) and is called the "genetic gain". On the right hand side, h^2 represents heritability mentioned in the previous chapter. And ΔP is called the selection differential, which is the difference in the quantitative trait between the group of individuals selected as parents and the group before selection (this formula is the quantitative expression of the relation between heritability and the effectiveness of artificial selection alluded to in Chap. 4).

Just as natural selection acts simultaneously on many traits of an individual, several traits of an individual are simultaneously considered in animal and plant breeding, and selection is often carried out to act as efficiently as possible in its entirety. Here the selection index (usually written as I) is generally expressed as a linear combination $I = b_1X_1 + b_2X_2 + \cdots + b_nX_n$ of the measurements of n traits, X_1, X_2, ..., X_n etc. (for example, egg weight, body weight, egg laying rate, etc. in

[3]Let me explain this with the haploid population model including genic selection used to derive formula (5.1). At one point in time (t) the genotypes A' and A exist in the population at frequencies p_t and $1 - p_t$ (in what follows, I omit the subscript t). If we denote the fitnesses of A' and A, expressed in Malthusian parameters, by m_1 and m_0, respectively, the selective value of A' is by definition $1 + s$ and hence $m_1 = \ln(1 + s) \approx s$, whereas for A we have $m_0 = \ln(1) = 0$. Thus, the mean fitness of the population is $\overline{m} = sp + 0 \times (1 - p = sp$, and the variance in fitness is $V_m = s^2p + 0^2 \times (1 - p) - \overline{m}^2 = s^2p(1 - p)$. The rate of increase per unit time of the mean fitness of the population is $\overline{m}$ differentiated by time t; this is $\frac{d\overline{m}}{dt} = s\frac{dp}{dt}$ and if we substitute the above formula (5b), $\frac{dp}{dt} = sp(1 - p)$, into the right hand side of this equation, this becomes $\frac{d\overline{m}}{dt} = s^2p(1 - p)$, which equals V_m. Thus, we obtain

$$\frac{d\overline{m}}{dt} = V_m \tag{5c}$$

In this calculation two alleles A' and A were assumed, but we can easily obtain the same result as formula (5c) for the haploid population model even if we assume the existence of three or more alleles A_1, A_2, A_3, ... But in higher organisms the diploid phase is developed, and it becomes necessary to use a population model of diploid individuals; in this case, not only the dominance relations and epistasis of genes, but also the form of mating (for example, whether inbreeding occurs) have an effect, and their complicated interactions make its mathematical treatment incomparably more difficult than the haploid model. In addition, simple expressions like formula (5c) or (5.2) are not generally applicable. Fisher avoids an important part of this difficulty by introducing a statistical concept called the "average effect" of the fitness of a gene, but based on my experience in dealing with detailed mathematical studies of this problem, it is my opinion that he goes too far in trying to claim that formula (5.2) is valid for biological organisms in general, and his discussions are perhaps farfetched.

chickens), and with the coefficients b_1, b_2, ..., b_n appropriately determined, the individuals are saved to be parents in descending order of this index. This is quite similar to the procedure used in an entrance examination in which the grade for each subject (corresponding to X) is appropriately weighted (corresponding to b), and students are accepted in order of their total grades (corresponding to I) until the places are filled. In the theory of the selection index it is an important question on what criteria the coefficients b_1, b_2, ..., b_n should be determined; Professor Yukio Yamada of Kyoto University has done excellent research on this (see reference at the end of this book).

Individual Differences in Fitness Are What Are Effective for Selection Darwin argued in *The Origin of Species* that organisms generally gave birth to many more offspring than could complete development; that the majority would die before maturity; that the struggle for life would be severe in nature; and therefore that natural selection would act effectively and organisms would evolve toward adaptation to the environment. However, as is clear from the "fundamental theorem of natural selection", what is effective for selection are the differences in fitness between individuals, and it cannot be said that selection will work only if each individual gives birth similarly to many offspring.

In this connection, Haldane makes the following point. In the oyster only about one among a million eggs can become a parent, whereas in the wild zebra an average of 1/2 to 1/3 of the 4–6 offspring that are born grow up to become parents. Nevertheless, the likelihood is much higher in the zebra for phenotypic change associated with genetic change to occur by selection. In other words, the fraction of mortality related to selection is likely much larger in the zebra than in the oyster (survival or death of oyster larvae is decided almost totally by chance). In fact, it has become clear in this century, by studies on natural selection in the field and on experimental populations, that almost all cases of natural selection usually observed entail the removal of deleterious variation and are not, as Darwin had in mind, associated with an increase of advantageous variation in the species.

Because of such developments, there are some science critics who have recently written in triumph that the logic leading to Darwin's theory of natural selection contains an error. However, it is important not to confound the long term possibilities (or inevitability) of selection with the current situation (or result); if Darwin on reading Malthus' theory of population recognized the severity of the struggle for life in nature and hit on the mechanism of adaptive evolution by natural selection as a general theory, it would be unjust to say that his thought processes were mistaken. Revolutionary theories in biology will, even if initially incomplete, be in many cases strengthened and refined with the passage of time if basically sound.

Classification of Natural Selection As already mentioned, if individuals of different genotypes are present in a population, and there are differences among them in the number of offspring contributed to the next generation (fitness), then natural selection can be said to be acting. However, strictly speaking, this should be called selection at the genotype level. To be more specific, natural selection acts in multifarious ways, and to clarify the whole picture for each species may be beyond the limitations of human intellect. To compensate for this, various methods of classification and their associated concepts have been proposed to obtain an overall and unified understanding of the workings of natural selection.

Positive Selection and Negative Selection We can first distinguish between positive selection and negative selection. Positive selection is the case in which a mutation appears in the population that improves viability or fertility (also called productivity), which gradually increases in frequency and spreads through the population, because individuals with this gene leave more descendants than other individuals. This selective action is fundamental to Darwin's theory of evolution, and it is appropriate to call it "Darwinian selection".

In contrast, negative selection is the case in which a deleterious gene appears in the population, the viability or fertility of an individual with this gene is adversely affected, and this gene is removed from the population. Negative selection is also called "purifying selection". Although positive selection is most important and fundamental for adaptive evolution, actual examples that have been clearly ascertained genetically are few—for example, the increase in the melanic form gene in the industrial melanism of insects, and the increase in the resistance genes associated with the continual use of pesticides, described in Chap. 2 and elsewhere—and in most cases are simply speculative. In contrast, many facts have been made clear regarding negative selection in connection with the essential nature of mutations. In particular, research on recessive lethal genes and slightly deleterious genes in fruit fly populations has recently provided us with many insights. I will discuss this subject in detail in the next chapter on population genetics.

Stabilizing Selection, Directional Selection, and Disruptive Selection However, it is in general not easy to demonstrate a clear connection between a genotype and selection, and in particular this is almost impossible for continuously varying quantitative characters such as height and body weight. Thus selection is often classified and treated at the phenotype level. The most common way is the trichotomous classification of selection according to K. Mather into stabilizing selection, directional selection, and disruptive selection.

"Stabilizing selection" is the form of natural selection believed to most commonly act on a quantitative character, and is the case in which individuals near the population mean have the highest fitness, and negative or positive deviations from this result in a decrease in the fitness of the individuals. An oft-quoted example is the relation between neonate body weight and subsequent mortality. According to research conducted in England by M. N. Karn and L. S. Penrose, mortality during the first month after birth was higher in the extremely light and extremely heavy babies compared to those of intermediate weights.

Stabilizing selection is, in a word, selection that removes extreme individuals; and for a normally distributed quantitative character such as in Fig. 5.2, death before reproduction of individuals corresponding to the shaded regions at both ends is of this category. The environment of biological species is circumscribed, and

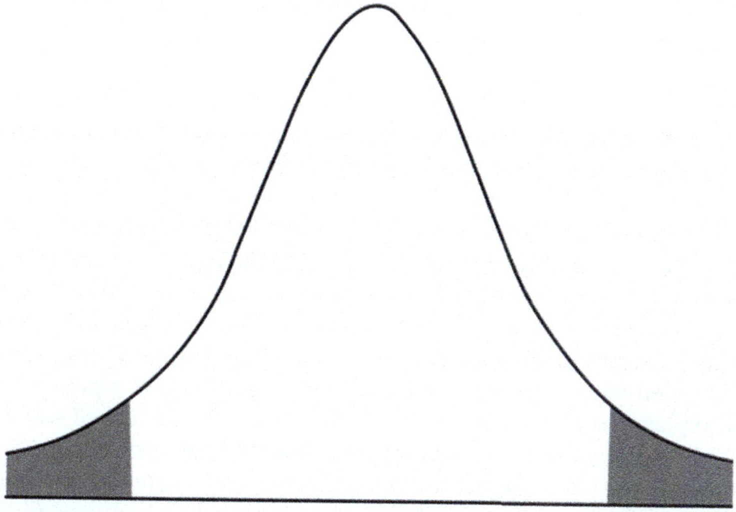

Fig. 5.2 An example of stabilizing selection. Shows a normally distributed quantitative trait, such that individuals at the two extremes in the shaded parts die before reproducing

phenotypically extreme individuals are likely often at a disadvantage because they fall outside the permissible range. In a different vein, we can argue that such individuals may be carrying many mutant genes, and that the genetic composition of a population is maintained every generation due to the balance between the increase in population variation by mutation and the removal by selection of both extremes. In any case, stabilizing selection has the role of maintaining the status quo.

Stabilizing selection is also referred to by the terms “normalizing selection” or “centripetal selection”. The observations made by Weldon on snails described in Chap. 2 is an example of this kind of selection in the wild, and several other similar observations have been reported.

Next, directional selection is selection that operates when the mean of a quantitative character is displaced from the optimum (the value with the highest fitness), and this causes the population mean to change in the direction of the optimum. If in nature the environment of a species suddenly changes, and the optimum that until then coincided with the mean moves elsewhere, directional selection will undoubtedly come into effect. This form of selection induces adaptive evolution.

The third, disruptive selection, occurs when there are two or more optima for one population in each generation. It is possible for disruptive selection to occur if the environment in which a population is placed is diverse with respect to natural selection, consisting of various ecological niches, and two or more phenotypes are each adapted to different ecological niches. At one time disruptive selection received exaggerated attention, and the claim was made that speciation could occur under its effect without geographical isolation, but this is now not generally supported. It should be added that when there are two kinds of alternative traits in the population and directional selection is reversed every generation, this is not called disruptive selection.

Index of Total Selection It must be noted with these forms of selection that, because there is selection at the phenotypic level, we cannot immediately say that selection is operating at the genotypic level. For example, as with individual variation in a pure line, if all variation between individuals is caused by environmental effects, then no matter how strong the selection applied to the phenotype, there is no effect on the next generation. On the other hand, if there are no differences between individuals in the number of children contributed to the next generation, there is no scope for selection to act at the genotypic level. Crow devised a quantity obtained by dividing the variance in the number of offspring per individual by the square of the mean and called this the "index of total selection". His purpose was to measure the potential (upper limit) for selection. When this index was calculated using data on births in the United States, interestingly, he obtained the result that the average number of children increased but that this index decreased between 1940 and 1960.

Frequency-Dependent Selection and Density-Dependent Selection In addition to the above, many terms with various adjectives appended to the word selection have been proposed. For example, in the field of "ecological genetics" which incorporates knowledge of ecology and investigates problems of adaptation genetically, the term "frequency-dependent selection" is often used. This refers to the case where the fitness of a genotype varies with its frequency; most important is a type of selection called minority advantage, in which a genotype is selectively advantageous when at low frequencies, but becomes disadvantageous at high frequencies. If a sort of division of labor exists between two alleles in the sense that different environmental resources are utilized, it is not surprising that such selection should operate. This kind of selection can be the cause of a balanced polymorphism in which two or more genotypes coexist in a population at stable equilibrium and are maintained at constant frequencies, and hence it was at one time made much of as a general mechanism for the maintenance of chromosomal and enzyme polymorphisms. At present, however, the view that such frequency-dependent selection is actually rare has gained support.

There is a similar term "density-dependent selection", which describes the case where the fitness of a genotype varies with population density. Common sense leads us to expect that the intensity of selection acting on different genotypes will increase where the density is high, but it is likely that in such cases also, what mostly happens is that selection that removes harmful variation (negative selection) will be intensified, and it will not necessarily often be the cause of an advantageous mutation spreading.

***r* Selection and *K* Selection** R. H. MacArthur and E. O. Wilson, in their treatment of the evolution of biological species on an island, proposed two forms of selection called "r selection" and "K selection". This concept was born in connection with the equation $dN/dt = rN(K - N)/K$, which describes the change in time of the number of individuals in a population (N). The left hand side of this formula expresses the growth rate of the number of individuals per unit time; r on the right hand side expresses the growth rate in the limit of low density, and K the maximum number of individuals that can be comfortably supported by environment. r and K selection have become very popular concepts nowadays in the field of ecology.

If there is sufficient food in the habitat, then while the density is low a genotype with a large intrinsic growth rate (r) has an advantage in selection (r selection). In this case, a genotype that consumes a sufficient amount of food, even if there is wastage, and produces many children is more adaptive than one that does not. However, when the number of individuals eventually increases and crowded conditions prevail, food per individual becomes scarce and under these conditions a different type of selection will operate. That is, a genotype with the ability to efficiently utilize even a small amount of food and to somehow raise its children becomes advantageous (K selection).

To determine whether or not these various kinds of selection are actually operating on biological species in nature is an important task for ecological genetics. Recently, J. A. Endler has reviewed and examined papers to date reporting that natural selection has been ascertained by direct observation in the wild, but the majority of these papers would seem to contain various problems, and satisfactory studies would seem to be few. From this we can infer the difficulty of proving natural selection.

The Creative Power of Natural Selection Lastly, let us think about the creative power of natural selection. Can an organism possessing a complex structural plan and with surprisingly advanced faculties like the human really be derived from a simple and lowly organism like the ameba, simply by the repetition of mutation and natural selection? This is a question that has been continuously raised against the theory of natural selection for nearly a century in the past, and is a question that remains even now in the minds of many people. As an even worse misunderstanding, orthodox evolutionary genetics is interpreted as if it were claiming that evolution from lower to higher can be explained merely by the accumulation of haphazard changes. This misunderstanding is not uncommon among the general public. It is generally not well understood that while the mutations per se are certainly without direction and as it were random, natural selection is the exact opposite and is a mechanism that produces order from disorder.

As described in the previous chapter, the information required to make the human organism is written with four types of DNA bases as letters on the approximately 3 billion base sites contained in the nucleus of a gamete. Let us now assume that a hundred million, corresponding to approximately 3% of these, are meaningful positions, and that the remainder are irrelevant to the expression of the human phenotype no matter which base is there (provided they are not deleted). Each of these hundred million sites is occupied by one of the four types of bases, and therefore if we ignore the fact that commands are scattered on 23 chromosomes, the commands for making a human correspond to one among the $4^{100,\ 000,\ 000}$ possible base sequences. Writing this number as N, we see that N is approximately

$10^{60,000,000}$, that is a number with 60 million digits. It is clear that this is an exceedingly large number from the fact that the total number of pairs of electrons and protons existing in the entire universe has been estimated to be about 10^{80}. The probability ($1/N$) that this base sequence would be produced by random mutation alone is $10^{-60,000,000}$, which is an extremely small number, and no matter how long a time and large a space is made available, it is realistically speaking impossible to be the result of chance alone.

On the other hand, if we assume that selection acted at each of the 100 million base sites, that advantageous variation accumulated step by step, and that this occurred during the past 3 billion years of the evolution of life on the Earth, (and if we consider that the appropriate base would by chance have already occupied approximately one fourth of the base sites), it would have been sufficient to substitute a base by natural selection (in the remaining parts) once every 40 years per site, and this is a plausible scenario.

Type Writing Monkey Let me explain this with another analogy, albeit on a smaller scale. If we were to have a monkey blindly type one page, and were to wait for a specific part of a Shakespearean play to be reproduced by chance, the probability would be almost zero even if we were to wait for a time corresponding to the beginning of the universe until the present (approximately 10 billion years). On the other hand, if we were to have the monkey type one letter at a time (corresponding to mutation), to erase it and have the monkey repeat what it did if it were not the correct letter, and to proceed if it were the correct letter; then, even if the adoption/rejection task (corresponding to natural selection) were to take about one minute, and it took almost an hour to successfully incorporate the correct letter, one year would suffice to complete the page (I will leave the calculation to the reader).

Muller's Explanation It was H. J. Muller who first clearly showed, in relation to the essential nature of mutation, the power of natural selection to change randomness into order and as it were to produce a miracle; already in 1929 he eloquently explained that, if the fraction of new alleles arising by mutation having an advantageous effect in survival over the previous gene is one in a hundred, and even if the total number of genetic loci is 1500, the probability of realizing the genetic composition of that organism purely by chance would be one in 100^{1500}, which is an exceedingly small probability that is impossible to realize in the actual world. By contrast, if natural selection is acting, there is no difficulty in accumulating 1500 mutations in the species by the evolutionary process.

Limit to Imagination The other impediment to understanding the creative power of natural selection is the limit to our imagination. Physics is able with the help of mathematical theory to argue about phenomena occurring in a world that greatly transcends, temporally and spatially, the range of daily experience, but such questions in biological evolution have not yet reached that stage. However, our imagination may be aided in this regard by the fact that splendid sculptures are carved from blocks of stone or wood by an artist (eventually the day will come when a computer with an advanced program will assist the sculptor) and are transformed into works of art that impress us.

God's Design or Natural Selection What is interesting in this connection is the following argument made by the eighteenth century theologian William Paley. Compared to an inanimate object like a stone in nature, a watch—the most precise instrument at that time—is much more complex and has an ordered structure, and it is clear that it could not have arisen by chance alone. In fact, this was designed and made by humans for a specific purpose. Similarly, the organs of biological organisms including the structure of the human eye have complex and purposeful functions, and must have been designed by someone. This someone is God, and it is averred that biological organisms are the strongest evidence demonstrating the existence of God.

Recently, R. Dawkins in a book entitled *The Blind Watchmaker* (1986) exhorts us in various ways that the explanation that biological organisms are designed by God is mistaken and that the complexity and purposefulness of the bodies of biological organisms can be satisfactorily understood as resulting from evolution under natural selection (I recommend this book to the interested reader). The blind watchmaker here is natural selection.

Introduction to Population Genetics 6

6.1 What Is Population Genetics?

Mendelian Population Genetics is usually regarded as a discipline dealing with how offspring resemble their parents, in other words how, with the individual as the unit of study, genes are transmitted from parents to offspring. In contrast, "population genetics" is slightly different, has the population rather than the individual as its object of study, and investigates the laws that control how its genetic structure is formed or changes. Of course, it is now one branch of genetics in the broad sense.

Moreover, the population that we deal with here is not a collection of heterospecific individuals but rather a collection of conspecific individuals that as a whole forms one breeding society, and is called a "Mendelian population" (all Japanese together form one Mendelian population). In this sense, it differs from the population in ecology. However, recently in the United States due to the influence of the late R. H. MacArthur, a tendency has arisen to regard population genetics and ecology from a unified evolutionary perspective, and the term "population biology" has recently come to be widely used.

Development and Its Consequences I will omit the developmental history of population genetics here, because I have already dealt with it in some detail in Chap. 2 in connection with the contributions of R. A. Fisher, J. B. S. Haldane, and S. Wright; in short, we can say that it was born when the efforts to unite Darwin's evolutionary theory based on natural selection and Mendel's laws of heredity using the methods of biostatistics came to fruition. The theory that forms the foundation of population genetics is mathematical in nature, and in the early stages of development the main subject was a so-called deterministic treatment, which ignores the random fluctuations of the proportions (frequencies) of various genes in the population. In this area, a series of studies by Haldane beginning in 1924 are particularly important. However, later, Fisher and Wright developed a theory that took into account the random fluctuations of gene frequencies, in particular the phenomenon of genetic drift in a finite population, as explained in Chap. 2. By these endeavors, the

M. Kimura, *My Thoughts on Biological Evolution*, Evolutionary Studies,
https://doi.org/10.1007/978-981-15-6165-8_6

theoretical structure of classical population genetics was mostly completed by the beginning of the 1930s.

The development of population genetics had the greatest impact on the theory of evolutionary mechanism, and it was already mentioned in Chap. 2 that the "synthetic theory of evolution", which eventually became the main stream in the academic circles of the world, took shape under its influence. In addition, the methods of population genetics were incorporated into the basic theory of animal and plant breeding, greatly contributing to its modernization. The theory of the selection index mentioned in the previous chapter is one example.

Of equal importance is the contribution to human genetics. In humans, where breeding experiments cannot be freely conducted, it goes without saying that statistical treatments of the population are indispensable to investigate the mode of inheritance, and the importance of population genetics in clarifying the genetic constitution of human populations will not change. Population genetics is relevant to the question of the frequencies in the Japanese population, not only of alleles involved in the normal range of individual variation as in height and blood groups, but also of deleterious genes causing various genetic diseases, and how they differ from other ethnic groups, and moreover what is the genetic character of Japanese. As a more practical question, knowledge of population genetics is absolutely necessary to estimate the genetically harmful effects on humans of environmental mutagens such as radiation.

Population Genetics at the Molecular Level About 20 years ago, population genetics experienced a major revolutionary change. The principal cause was that it became possible due to the development of molecular biology to analyze genetic variation within a population at the molecular level, i.e., at the level of the internal structure of the gene. As a result, it became clear that much more genetic variation was present in a population than generally thought until then. And a heated controversy arose surrounding the mechanism(s) of its maintenance, in which a position close to traditional panselectionism and the position of the neutral theory that attaches importance to random drift of selectively neutral mutant genes were opposed. In addition, it became possible to deal with the question of biological evolution at the molecular level, and, as already explained in Chap. 2, a controversy arose here between the neutral theory and selection theory.

What is important for population genetics is that a new field was pioneered called population genetics at the molecular level. In particular, progress was made with a mathematical method called "diffusion models", which treats gene frequency change as a continuous stochastic process, i.e., as probabilistic events that proceed in continuous time, and large advances were made that finally transcended the framework of classical theory constructed in the 1930s by Fisher, Haldane, and Wright (I will explain the neutral theory in Chap. 8).

With regard to experimental aspects, the fruit fly has long played a major role in the genetic studies of natural populations. In particular, from about 20 years ago, genetic variation of proteins and especially enzymes began to be investigated in natural populations of the fruit fly using electrophoresis (a method in which an electric field is applied to a gel medium to detect differences in the mobility of proteins relative to the electrodes; and it was found that in populations of the fly there were unexpectedly many protein polymorphisms in which two or more alternative variants both existed at high frequencies, and their importance as research material increased further. Moreover, due to the possibility of precisely measuring fitness, and through selection experiments in population cages, the fruit fly became increasingly useful as research material. Subsequently, with the introduction of research on DNA polymorphism, the current situation has become such that population genetics cannot be debated without a comprehensive understanding of the results obtained from studies on the fruit fly.

In addition, it is a characteristic of recent research that a wide range of biological species from *E. coli* to the human are being used in the study of protein and DNA polymorphisms. Furthermore, the importance of human populations as research material is increasing. Humans may be the most appropriate material for research on the breeding structure of populations, including the migration of individuals between regional populations.

The Role of the Computer We also cannot ignore the role of the computer. Here, what is important is not the processing of large amounts of data as is usually said, but simulations conducted on the computer. It is now possible to have the computer simulate a sexually reproducing population and to investigate with relative ease how its genetic structure changes in each generation under various hypothetical evolutionary forces. The so-called Monte Carlo experiments are extremely useful, in which simulations that include probabilistic events such as mutation and genetic drift are conducted by generating random numbers. The computer has now become an indispensable tool for so-to-speak experimentally investigating, by the Monte Carlo method, the problems that are too complex and do not permit a clear and neat mathematical treatment, or those cases in which only an approximate solution can be obtained analytically, and to obtain by applying large scale numerical methods the answers to difficult equations that are virtually impossible to deal with analytically.

6.2 Gene Frequency and Mating System

Why Gene Frequency? The most basic quantity that expresses the genetic composition of a population is the proportion of each type of gene in the population, i.e., the "gene frequency". Higher organisms including the human have a well-developed diploid phase with two sets of chromosomes, and the diploid individual is the object of study. Notwithstanding, it is not the actually observed genotype of an individual (in the case of ABO blood groups, an individual of type AB for example), but rather the indirectly estimated gene frequency (for example the frequency of gene A) that is regarded as fundamental for the following reason.

If we consider a sexually reproducing population, it usually contains an enormous number of genotypes. For example, at a locus where two alleles A and a are segregating, three genotypes AA, Aa, aa may occur, and if there are 100 such loci,

3^{100} in other words approximately 5.1×10^{47} different genotypes are possible. Even if a species comprising one million individuals in each generation were to persist for one million generations, the total number of individuals would be 10^{12}, and the various genotypes actually appearing in the population would be only a small subset of the possible combinations. In fact the total number of segregating genetic loci is not of the order of 100, but may be several thousands. As such, in the evolutionary history of the human, the probability is virtually zero that individuals with the same genotype will independently arise except in the case of monozygous twins. Thus, the genotype does not possess continuity in transmission from parent to offspring. By contrast, the gene itself is transmitted from parent to offspring by a self-replicating action, and hence the proportions (frequencies) of the various genes in the population are quantities that change relatively slowly with time. This also tells us that the frequencies of genes are quantities that can be theoretically treated much more easily than the frequencies of the genotypes that are formed by their combination. For any particular genetic locus, the gene frequency can be obtained by adding the frequency of the homozygote for that gene and one half of the frequency (frequencies) of the heterozygote(s).

Mating System The first step in the theory of population genetics is to focus on one genetic locus and to examine gene frequency change at that locus. A problem that immediately arises is that natural selection does not act directly on the gene, but rather generally acts at the level of the individual on the various genotypes; and that it is therefore insufficient to be given the frequencies of genes, but is generally also necessary to know the frequencies of genotypes. For this, it is necessary to stipulate the mating system that determines how the homologous genes that come from the parents combine to produce the individual.

The Assumption of Random Mating and the Hardy-Weinberg Principle Here, the simplest (and moreover useful) conceivable mating system is "random mating". Roughly speaking, it is a mating system in which two homologous genes are non-preferentially and randomly combined to produce an individual. This can be thought of as corresponding to the situation where matings occur between male and female individuals in the population without preference and at random. The assumption of random mating is useful because when this kind of mating occurs an extremely simple relation holds between gene frequencies and genotype frequencies, i.e. the "Hardy-Weinberg principle".

Let us now consider the case in which in a sufficiently large (ideally an infinite) randomly mating population, alleles A_1 and A_2 exist at frequencies p and q $(=1-p)$, respectively, at an autosomal genetic locus. From the assumption of random mating, the probability (frequency) that the same allele A_1 will be transmitted through both male and female gametes, and that the two will combine to form the homozygote A_1A_1 is the square of p, i.e., p^2. Similarly, the probability that A_2A_2 will be formed is the square of the frequency of A_2, i.e., q^2; and moreover the probability that the heterozygote A_1A_2 will be formed is $2pq$, because A_1 may come from the male gamete and A_2 from female gamete, or vice versa. This is often expressed as

$$(pA_1 + qA_2)^2 = p^2A_1A_1 + 2pqA_1A_2 + q^2A_2A_2.$$

In this formula, $pA_1 + qA_2$ within the parentheses on the left hand side is sometimes called the gametic array, and $p^2A_1A_1 + \cdots$ on the right hand side the zygotic array. Moreover, in deriving this formula it is assumed that the gene frequencies are identical in males and females; but even if there is a difference, so long as the gene is not on the sex chromosome, in the population of offspring born in the next generation there will be no difference between males and females, so that the difference in frequencies will disappear in one generation. In addition, if factors that cause gene frequency change are absent, such as mutation, selection, and exchange of individuals with other populations, then p and q are invariant, and this formula holds in every generation for the same p and q. This is the gist of the Hardy-Weinberg principle.

As is clear from this explanation, this principle can be readily extended to the case of more than two alleles; if $A_1, A_2, A_3, \ldots$ exist at frequencies $p_1, p_2, p_3, \ldots$, the zygote frequencies can be obtained from $(p_1A_1 + p_2A_2 + p_3A_3 + \cdots)^2 = p^2A_1A_1 + \cdots$.

Moreover, the assumption of random mating is a kind of theoretical simplification, but in bisexually reproducing animals predictions based on this assumption in many cases agree well with actual observations and is therefore useful. For example, with regard to the MN blood groups, a survey of approximately 50,000 Japanese has yielded data showing that the observed frequencies of *MM*, *MN*, *NN* are 29.599%, 49.460%, 20.941%, respectively; when the frequency of the M gene is obtained by adding half the frequency of *MN* to the frequency of *MM*, it comes out to 0.54329 (let this be p), and when we take its square (p^2) it is 0.29516. Therefore, this theoretical value for type *MM* is in excellent agreement with the observed frequency (0.29599).

Useful But Simple The Hardy-Weinberg principle is useful but theoretically simple, and it is not necessary as is done in many textbooks to explain it pompously as something difficult. By the way, this principle is traditionally stated in the following way.

1. In a population where factors such as natural selection, mutation, migration, genetic drift are absent and random mating occurs, the frequencies of the various genotypes at a genetic locus on an autosome do not change over generations.
2. If alleles A and a are present in the proportions (relative frequencies) p and q, respectively, we have under random mating

$$p^2AA : 2pqAa : q^2aa$$

(note $p + q = 1$), and these proportions are subsequently maintained over the generations under random mating.

This is accompanied by a footnote saying that of the two (i) is more important. There is a historical reason why (i) is viewed as more important; around the beginning of this century, there were some biometricians who did not fully understand the population genetic implications of Mendel's laws, arguing that if Mendel's laws were correct, dominant traits (for example brachydactyly in humans) would increase in frequency in the population ultimately reaching 75%; and because such a thing was not possible, offering misplaced criticisms such as that Mendel's laws were mistaken. Hardy's paper (1908) played an important role in correcting this misunderstanding. Now that the explanation of Mendel's laws based on the behavior of chromosomes has been established, such a misunderstanding no longer exists. What is still meaningful is (ii), which is the convenient property that in a random mating population, the frequencies of genotypes can be obtained by suitably multiplying the frequencies of alleles.

A Ridiculous Tendency The Hardy-Weinberg principle is nowadays covered even in high school biology textbooks, and this is what students of population genetics first learn. However, it is in my opinion ridiculous that in many cases (i) is treated as more fundamental and that Hardy-Weinberg principle is stated to be the basis of population genetics.

I do not know where this tendency originated, but it may be from the famous book *Genetics and the Origin of Species* third edition (1951) by T. Dobzhansky. In this third edition Dobzhansky writes exaggeratedly that this principle is "the foundation of population genetics and modern evolution theory". In my opinion, to teach the Hardy-Weinberg principle in this way is truly futile and moreover a teaching method that will confuse students. In all probability, the writings of Dobzhansky and other American scholars who did not fully understand the theory of population genetics were regarded as exemplary, and then copied by authors of textbooks in the United States and Japan, resulting in this state of affairs.

In retrospect, in the absence of specific factors (mutation, segregation distorters, natural selection, etc.) that cause gene frequency change, gene frequency change does not occur by Mendelian segregation alone, and hence what is stated in (i) is obvious and does not require mathematical proof. This is also evident from the fact that a gene is a self-replicating molecule. The formation of a zygote entails only that two homologous genes are combined; and although importance is attached to the frequency of zygotes because the diploid phase is well developed in the higher organisms, I doubt that anyone would take notice of the Hardy-Weinberg principle if the human were a haploid organism.

Rather, the meaningful part of this principle even now is (ii), and even when gene frequencies change over generations by natural selection, what is extremely useful is that the zygote frequencies immediately after fertilization under random mating are given approximately every generation as the product of the gene frequencies among the parents at that time (however, caution is required, because if viability selection acts during development, this does not hold after selection without an appropriate correction, and it is also not appropriate when there is selection on fertility).

Inbreeding and Assortative Mating The assumption of random mating is useful, but does not hold in every case. The especially important cases of non-random mating are "inbreeding" and "assortative mating". Inbreeding is mating between consanguineous individuals, i.e., mating between close relatives. On the other hand, assortative mating is the case in which phenotypically similar individuals are more likely to mate; in humans it is said that height and intelligence quotient are positively correlated to a certain extent between husband and wife, and hence it is possible that assortative mating occurs for these traits.

Generally speaking, inbreeding is biologically much more important than assortative mating, and the genetic consequences that arise from this are problems that cannot be avoided in human genetics and in breeding. Fortunately, the method of the inbreeding coefficient developed by S. Wright and the French mathematician G. Malécot is effective, and many problems can be satisfactorily dealt with by this method.

Although I cannot address this problem in any depth in this book, I would like to say a word or two about the inbreeding coefficient. This is the probability that homologous genes in an individual are descended from a common ancestral gene. The inbreeding coefficient is usually denoted by the symbol F. As an example, if we consider an individual (A) born between unrelated parents, $F = 0$ for this individual. If this individual is a plant and is self-fertilized, an offspring born from this individual has an inbreeding coefficient of 1/2. The reason for this is that the homologous genes in this offspring are derived from either one of the two homologous genes in parent A with probability 1/2. Although the calculation is more involved, it can be shown taking the human as an example that a child born of a cousin marriage has an inbreeding coefficient of 1/16. In addition, although I will omit the explanation, in a collection of individuals with a fixed value F of the inbreeding coefficient, the frequency of the homozygote aa is not q^2 as is the case in a random mating population ($F = 0$), but $(1 - F)q^2 + Fq$. If a is recessive gene causing a rare genetic disease and its frequency in the population is 1%, the proportion of children with this recessive genetic disease born between unrelated parents is 1 in 10,000, whereas from a cousin marriage the proportion born is approximately 7 times this number.

6.3 On Genetic Equilibrium

Genetic Equilibrium As already mentioned, the theory of population genetics was at first a deterministic one that ignored chance fluctuations in gene frequencies. This is much simpler than the theory as a stochastic process that developed later. Nevertheless, this provides sufficiently accurate predictions and is extremely useful in the case where natural selection is strong and the population is large. Thus, it is widely used even now. In the previous chapter, assuming "genic selection", I explained the calculation method for obtaining the number of generations required for a mutant gene that is advantageous for survival to increase from a low frequency (for example 0.1%) to a high frequency (for example 99.9%), and this is typical of the deterministic approach.

However, in general, a mutation advantageous for survival rarely occurs, and the majority are deleterious. Genes that cause genetic diseases in the human have arisen as a result of this kind of mutation. Such mutations, of course, cannot spread in the population, but rather tend to be removed by selection. However, they are not entirely lost from the population, because they are continuously generated by mutation and replenished in the population. In this section, I would like to explain the genetic equilibrium that is maintained by the balance between mutation and negative selection.

Achondroplasia A well-known genetic disease is achondroplasia (short limbs). This is due to a dominant mutant gene, and a child who inherits this gene from either the father or the mother has short limbs. This is a disease in which height is extremely low, but it is known that a patient who has survived beyond a certain developmental stage is healthy and can live for a fairly long time. In this example, the following formula holds when mutation and selection are balanced in the population.

$$2v = x(1 - w) \tag{6.1}$$

Here, v on the left hand side is the mutation rate, i.e., the rate at which the mutant gene for achondroplasia arises every generation. This is approximately of the order of several in 100,000. x is the frequency of patients, and is a quantity that expresses the proportion of babies that have achondroplasia at birth. And w is the relative fitness of a patient; taking into account the entire process in which a baby with achondroplasia grows up, marries, and gives birth to children, this is the rate at which children of the next generation are produced relative to the normal person.

The meaning of this formula is as follows. The left hand side is the rate at which patients with achondroplasia are added to the population every generation by mutation, and it is multiplied by 2 because the mutation can descend from either parent. On the right hand side, w denotes the rate at which children are produced, so $1 - w$ is the proportion removed by natural selection, and $x(1 - w)$ is the quantity representing the loss of the patients every generation from the population. Then, equilibrium is attained where this decrease is balanced by the increase corresponding to $2v$.

This argument was first formalized by J. B. S. Haldane. The value of x in this formula can be estimated by counting the number of patients among babies. In studies conducted in Demark, one such patient is born per approximately 9400 births. Then, when the patient is followed and the number of children produced relative to the normal person is investigated, only about one fifth is produced compared to the normal person. Hence, when $x = 1/9400$ and $w = 1/5$ are substituted into this formula and the calculation is done, the mutation rate v comes out to 4.3×10^{-5}. To sum up, the gene for this disease arises every generation from the normal gene at a rate of 4.3 per 100,000.

Whether or not this is correct can be ascertained by a different survey. That is, we can investigate the cases in which both parents are normal and the children born have achondroplasia. In a survey conducted again in Demark, among 132,761 births from normal parents 11 babies with achondroplasia were born. The gene can mutate when transmitted from either parent, hence 11/132, 76111 corresponds to $2v$, and dividing this by 2 yields the answer $v = 4.1 \times 10^{-5}$. This agrees well with the previously obtained value of $v = 4.3 \times 10^{-5}$. Therefore, genetic equilibrium can be said to hold, and deterministic theory is shown to produce a consistent result, albeit approximately.

In this way, indirect estimation of the mutation rate assuming genetic equilibrium is one valuable research means in human genetics.

Recessive Lethal Genes in the Fruit Fly As my second example, I would like to consider recessive lethal genes in the fruit fly. When individuals are collected from wild populations of *Drosophila melanogaster*, and the second—one of several—chromosomes of the fly are studied, about 20% of them contain a recessive lethal gene somewhere along the chromosome. A recessive lethal gene is a deleterious gene that causes death of the individual in the homozygous state, and it is believed from other studies that such mutations can occur at approximately 500 genetic loci on the second chromosome. Therefore, there are many types of recessive lethal genes, any one of which when homozygous expresses a lethal effect at various stages in the development of the fly. But in the heterozygote, i.e., when the fly has inherited such a gene from only one parent and the normal allele from the other parent, viability is almost normal.

However, I would like to draw your attention to the fact that these so-called recessive lethal genes are not completely recessive with regard to their deleterious effects, but lower the viability of the fly by a very small amount in the heterozygote, i.e., between about 2% and 5%. This possibility was first pointed out by H. J. Muller and A. H. Sturtevant, and received strong opposition at one time from Dobzhansky and his school, but was later convincingly shown to be correct by the large scale studies of J. F. Crow, Professor Terumi Mukai, and others.

The rate at which recessive lethal genes arise by mutation has been also studied in experiments since the days of Muller, and for the second chromosome of *Drosophila melanogaster* the spontaneous mutation rate to lethal genes is approximately 0.5% per chromosome per generation. If the total number of genetic loci involved is 500 as already mentioned, the lethal mutation rate per genetic locus is approximately 10^{-5}.

Equilibrium Frequency of a Lethal Gene Let us now focus on one of these genetic loci and denote the normal gene by + and the recessive lethal gene that arises from this by l (l is the abbreviation of the word lethal). In general, let v be the rate at which l arises from + by mutation per generation, and let the viability of the heterozygote $+/l$ be $1 - h$ relative to the viability of the homozygote +/+ which we set to 1. Here h is the selection coefficient that expresses the strength of negative selection against the individuals that are heterozygous for the lethal gene.

Next, let us introduce the idea of genetic equilibrium and consider the state at which a balance is maintained between the de novo generation of mutations and their loss by selection. Then, if we denote the equilibrium frequency of the lethal gene l by $\widehat{q}$, the following formula holds:

$$v = \widehat{q}h. \tag{6.2}$$

The left hand side of this formula represents the increase by mutation, and the right hand side the decrease by negative selection. This formula is derived by noting that the two quantities are equal at equilibrium. However, this formula only takes into account selection on the heterozygote and ignores selection due to lethality of the homozygote. The reason for this is that we have assumed that homozygotes rarely occur because l is at low frequency (amounting to $\widehat{q}^2$) and that the result is unaffected when we ignore the existence of l/l (this can be checked after obtaining the gene frequency $\widehat{q}$).

This formula can also be expressed as

$$\widehat{q} = \frac{v}{h}. \tag{6.3}$$

According to the laboratory studies by Professor Terumi Mukai and his colleagues on the viability of heterozygotes, the value of h averages about 2%. Thus, setting $h = 0.02$ and using the mutation rate $v = 10^{-5}$ obtained above, we obtain for the equilibrium frequency, $\widehat{q} = 5 \times 10^{-4}$, from formula (6.3). Because we have assumed that lethal genes may arise at 500 loci on the second chromosome, the average frequency per second chromosome is 500 times this amount, i.e., 25%. This agrees well with the observed value of approximately 20% obtained from surveys of natural populations (if we correct for the case where two or more recessive lethal genes are by chance found on one chromosome, the theoretical value becomes 22%, in even better agreement with the observed value).

Distribution of Deleterious Effects Because the second chromosome constitutes about two fifths of the total chromosomal complement of the fruit fly in size, the average number of recessive lethal genes per haploid chromosomal set is approximately 0.5, and hence, because the fly has two sets of chromosomes, it carries an average of approximately one recessive lethal gene per individual in the heterozygous state. Of course, this is for apparently normal individuals collected from natural populations, and it is impossible to tell from outward appearance that they harbor harmful things such as lethal genes. In addition, wild flies that look healthy carry recessive deleterious genes that are not lethal but lower viability to various degrees in homozygotes. Recessive semi-lethal genes are one example.

Here, the frequency distribution of the deleterious effects of genes that arise by mutation is of interest. If we collect many individuals from a natural population, take many of the sampled second chromosomes, make them homozygous, and examine the effect on the viability of the individuals, there are two frequency peaks; one peak corresponds to the group of chromosomes with viabilities that are indistinguishable from the normal, and the other peak corresponds to the group of lethals (for the sake of convenience, chromosomes with viabilities less than 5% are placed in this group); and there are very few that lie in the intermediate range (chromosomes with viabilities between 5% and 60% in homozygotes). From this, we can see that the distribution of deleterious effects deviates hugely from the uniform distribution, the majority are recessive lethals, and semi-lethals with moderate deleterious effects do not often arise. However, according to studies by Professor Mukai, there is evidence to suggest that "slightly deleterious genes" (Mukai's "viability polygenes"), which lower viability very slightly (an average of between about 2% and 3%) in homozygotes, have a mutation rate about 20 times higher than lethal genes.

Another notable thing is the degree of dominance of "recessive" deleterious genes. This is that the magnitude of the deleterious effect in the heterozygous state is about the same for genes that are lethal, semi-lethal, or slightly less harmful in homozygotes (about 3%), and that there is little difference (there is little association between the deleterious effects in heterozygotes and in homozygotes). It is also known that deleterious genes that arise by mutation do not always lower viability, but some recessive genes are present in natural populations that cause sterility.

Even Healthy Persons Carry Several Lethal Genes Such results appear at first sight to have no bearing on our daily lives, but this is not necessarily so. Humans are no different from flies in that they harbor deleterious genes such as recessive lethal genes in the heterozygous state, and in this respect the human can be viewed as a large fly. Of course, mating experiments cannot be freely conducted on humans, but by investigating the increase in the rates of abortions and still births and in mortality rates during infancy for consanguineous marriages (for example cousin marriages), we can estimate, albeit indirectly, the number of recessive lethal genes per individual. The study by N. Morton, J. F. Crow and H. J. Muller (1956) is well-known; they obtained the result that humans harbor an average of between 1.5 and 2.5 lethal equivalents of recessive lethal genes per gamete (twice this number per individual). Here a lethal equivalent is the number of deleterious genes converted to lethal genes, for example, two genes causing semi-lethality are the equivalent of one lethal gene. It is difficult to imagine on common sense that, even as we live our daily lives in apparent health, we harbor between 3 and 5 recessive lethal equivalents of deleterious genes in the heterozygous state, and that if we were suddenly to become homozygotes, we might on average die about two times.

As already mentioned, "recessive" deleterious genes are not completely recessive in the fly, but manifest a very small deleterious effect in the heterozygous state. Although unconfirmed, the same may be true of the human, and it is possible that a significant part of the idiosyncratic health problems (taking after our parents)—it may be inappropriate to call them genetic diseases—may result from such heterozygosity.

H. J. Muller in 1950 in a paper entitled "Our load of mutations" rejected the view prevalent at that time among medical doctors that mutations were insignificant as the cause of diseases in mankind, and pointed out for the first time that they could be a major cause of death from birth to reproductive age and of sterility. In this paper Muller uses the term "load" in the usual sense of a burden. In other words, it expresses the various handicaps in life that are suffered by individuals who carry the deleterious genes that have accumulated in human populations by the mutations occurring every generation.

Genetic Load Later, Crow extended Muller's idea and proposed the concept of "genetic load" (1958). This expresses the magnitude of natural selection acting at the level of the genotype, and according to his definition, genetic load expresses the proportional reduction of the mean fitness (mean selective value) of the population relative to the optimal genotype in the population. In other words, genetic load (L) is given by the following formula

$$L = \frac{\left(w_{op} - \overline{w}\right)}{w_{op}} \tag{6.4}$$

Here w_{op} is the selective value of the optimal genotype and $\overline{w}$ is the mean selective value of the population. Let us apply this to the recessive lethal gene discussed above.

If we consider one genetic locus, and let the mutation rate from the normal wild type gene (+) to the recessive lethal gene (l) be v, the frequency of l at equilibrium, $\widehat{q}$, is given by formula (6.3). In this case, the optimal genotype is the normal homozygote +/+, and taking its fitness as the standard we have $w_{op} = 1$, and the fitness of $+/l$ is $1 - h$. Furthermore, l/l is lethal, so that its fitness is 0.The frequencies of the three genotypes are $(1-\widehat{q})^2$, $2\widehat{q}(1-\widehat{q})$, $\widehat{q}^2$, and if we can safely ignore $\widehat{q}^2$ because the value of $\widehat{q}$ is small, the mean selective value of the population will be $\overline{w} = 1 \times (1-\widehat{q})^2 + (1-h)2\widehat{q}(1-\widehat{q}) \approx 1 - 2h\widehat{q}$, and on substitution of formula (6.3) into this we obtain $\overline{w} = 1 - 2v$. Therefore, from formula (6.4) we obtain the simple result

$$L = 2v. \tag{6.5}$$

In words, the genetic load due to recessive lethal genes that have a deleterious effect in heterozygotes, i.e., the reduction of the population fitness, is equal to two times the mutation rate per gamete.

Here, it is noteworthy that this formula holds independently of the deleterious effect h in the heterozygote. However, this formula does not hold if $h = 0$, and it is a necessary condition that h is much larger than v. When this condition is satisfied, the mutation need not be lethal in the homozygote, and it is sufficient that it has at least two times as deleterious an effect in the homozygote as in the heterozygote. In addition, the right hand side of formula (6.5) corresponds to the mutation rate per individual at this genetic locus, and this can be viewed as being equal to [the contribution of this locus to] the mutational load. This is quite reasonable if we think about it, as equilibrium will not hold unless the mutations arising per individual and the number removed by the death of an individual are the same.

Haldane-Muller Principle The properties of the mutational load expressed by formula (6.5) are sometimes called the "Haldane-Muller principle". The importance of this principle becomes more apparent when we consider the totality of genes in an individual rather than the individual genetic loci.

The total number of deleterious mutations that arise per individual every generation is not clearly known, but if we assume that they occur at 30,000 genetic loci at an average rate of $v = 10^{-5}$, and that the deleterious effects of the different mutations are independent and mostly additive, then the total mutational load will be 0.6. That is, 60% of the offspring that are born will die of genetic causes before completing growth. When we add environmental (non-genetic) deaths, the mortality rate will be even higher.

In primitive human societies before farming, it is not clear how many children were born to parents; I once read that in the primitive state the standard condition was for an average of four children to be born, of which half would complete development. If this is so, a mutational load greater than 0.5 would be unbearable, and hence the total mutation rate to deleterious genes may be much lower than 0.6. An alternative consideration is the possibility that there is reinforcement of the deleterious effects between deleterious genes at different loci (a kind of epistasis), and that as the number of deleterious genes per individual increases the deleterious effect grows in proportion to a power of that number. In this case, it has been proven mathematically that the total load is reduced to a fraction of the total mutation rate (roughly speaking, the load is reduced because, in this case, two or more deleterious genes are simultaneously removed from the population by the death of one individual). Moreover, mutations that are neutral with regard to selection do not contribute to the load, and hence they obviously are inconsequential no matter how many occur.

6.4 On Genetic Drift

Random Fluctuations of Gene Frequency In natural populations, when sexual reproduction occurs, many gametes are produced, but only a comparatively small number actually contribute to forming the next generation. I have already mentioned in Chap. 2 the random fluctuations of gene frequency that occur because of this, i.e., genetic drift; nevertheless I would like to explain this here in a little detail, because the concept of drift forms the basis of the neural theory of molecular evolution that will be dealt with in Chap. 8. In what follows my discussion will assume alleles A_1 and A_2 that are neutral with respect to natural selection.

Let us consider a diploid population comprising a fixed number, N, of individuals per generation and assume that sexual reproduction proceeds in the following way. First, it is assumed that a gene pool comprising an infinite number of gametes is produced, then N male gametes and N female gametes are randomly sampled from this and combined to form the N individuals of next generation. This is an abstract representation of a population in which matings occur without mutual preference and including self-fertilization (random mating population), and if such a process of reproduction is iterated, the frequencies of the alleles gradually change by chance. This is as it were similar to repeating an operation in which black balls and white balls are put into a big bag and a fixed number of balls are blindly picked out of this bag.

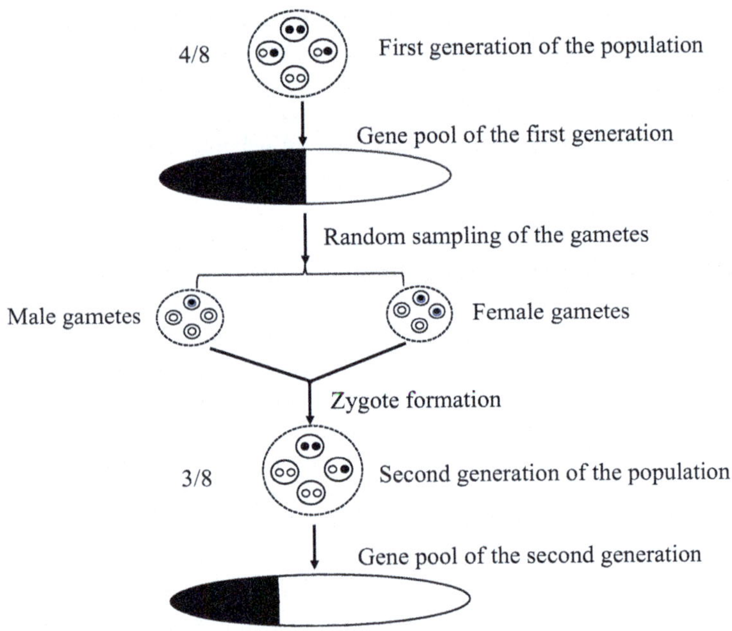

Fig. 6.1 Random sampling and random fluctuation of the gene frequency in a virtual population comprising N = 4 gametes

Figure 6.1 shows the case of a population comprising four breeding individuals ($N = 4$), and alleles A_1 and A_2 are represented by black circles and white circles. In this example, the frequency of the black circle gene (A_1) in the first generation is 4 in 8, i.e., 50%. In the transition to the next generation, the black circle gene has entered exactly one half of the four female gametes sampled from the gene pool, but it has entered only one of the four male gametes; as a result, the frequency of black circle gene in the population of the second generation has decreased to three in eight, i.e., 37.5%. In this example, the population is exceptionally small, consisting as it does of four individuals, and hence the frequency changes rapidly; actual natural populations usually comprise a minimum of several thousand individuals and are not rarely of the order of 100,000. Therefore, the magnitude of change per generation is expected to be much smaller than this.

In general, in a population comprising N breeding individuals, setting the frequency of A_1 to p, it easily follows from a calculation involving the binomial distribution, which is well known in probability theory, that the standard deviation of the magnitude of change per generation due to drift is $\sqrt{p(1-p)/2N}$. Hence, if the gene frequency is 50% ($p = 0.5$), and the number of breeding individuals in the population is 10,000 ($N = 10^4$), the standard deviation of the magnitude of change per generation will be the small value, 0.0035. However, such change is cumulative,

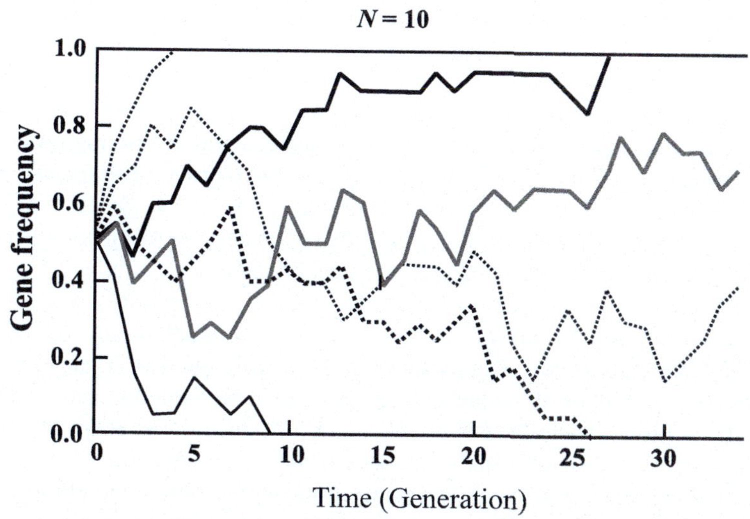

Fig. 6.2 The results of a Monte Carlo experiment conducted on a computer, assuming a population comprising 10 breeding individuals in each generation, to investigate how the frequency of allele changes as the generations elapse

and after 20,000 generations in this example, the gene frequency in the population will have changed completely.

Results of Computer Experiments Figure 6.2 shows an actual example of how the gene frequency in a population changes with time (generations) by genetic drift. That is, it illustrates a part of the results of replicated Monte Carlo experiments in which the computer was made to simulate a sexually reproducing population comprising ten individuals and the gene frequency was started from 50%. Here, each irregular line shows the process of change in the frequency of A_1 over many generations for one population. As can be seen from the figure, there are cases in which A_1 is luckily fixed in the population within several generations (the irregular line at the very top of the figure), and there are also cases in which it is unluckily lost from the population before ten generations have elapsed (the irregular line at the very bottom). In the majority of cases, it lies between the two; but it can be proved mathematically that it will not continue to fluctuate up and down indefinitely, and that it will eventually end in fixation or loss. Furthermore, for the population of ten individuals illustrated in this figure, it can be proved that the average time till fixation or loss starting from a frequency of 50% is about 28 generations.

Inbreeding Effect One important property of genetic drift in a finite population is that descendent genes derived by chance from a common ancestral gene accumulate in the population, and hence that a kind of inbreeding effect emerges, and genetic variation in the population decreases (in humans there were formerly many instances in which everyone in an isolated village became related, which is due to a similar principle). As already mentioned in Chap. 2, Wright showed for the first time that, in a population of N breeding individuals, the rate of decrease of variation per generation by drift was $1/(2N)$. Moreover, using the concept of the inbreeding coefficient (F) explained in Sect. 2 of this chapter, it can be proven by a simple calculation that the value of F increases at the rate $1/(2N)$ every generation (the interested reader should refer to Chap. 2 of Iwanami Lectures Current Biological Science 6 *Fundamentals of Human Genetics*). Therefore, unless new genetic variation is supplied by mutation, the population gradually becomes genetically homogeneous at this rate (that is, the frequency of homozygotes increases).

Effective Population Number I would like to draw your attention to the fact that it is not the apparent number of individuals in the population that is relevant to drift, but rather the number of individuals that participate in reproduction, and to deal with this problem more generally it is necessary to introduce the concept of "effective population number". This is usually denoted by the symbol N_e, and various formulas have been contrived for N_e, such that populations with different breeding structures would always have a rate of decrease of genetic variation of $1/(2N_e)$ every generation. Examples are when the number of breeding individuals among males and females differ greatly or when the number of individuals in the population periodically declines drastically but reverts to the original size a few generations later. In such cases, the effective population number (N_e) will be much smaller than the apparent size (N).

6.5 Behavior of a Mutant Gene in the Population

At First, Chance Prevails Let us assume that a new allele (call this A') has appeared in the population by mutation. Because a mutation is almost always unique at the molecular level, it usually exists in single copy when it appears in the population. And if initially there is only one copy, it will be carried in the heterozygous state by one individual. The fate of such a mutant gene that exists in single copy will, even if the population is large, be determined almost completely by chance during the first few generations. This is due to the random sampling of gametes during sexual reproduction, and is clear from the following considerations.

Among the many gametes produced by the individual that is heterozygous for A', the frequency of A' will be 50%; and if this individual contributes two offspring to the next generation, only two gametes can be transmitted to the next generation, and the probability that neither of the two carries A' is $(1/2)^2$. In other words, A' is completely lost from the population with probability 1/4.Moreover, even if it survives in the next generation, it may be lost in the generation after next, and the calculation becomes complicated as the generations are repeated. The general treatment of this kind of problem requires a specialized knowledge of probability theory, so I will omit it and confine myself to noting two or three important results.

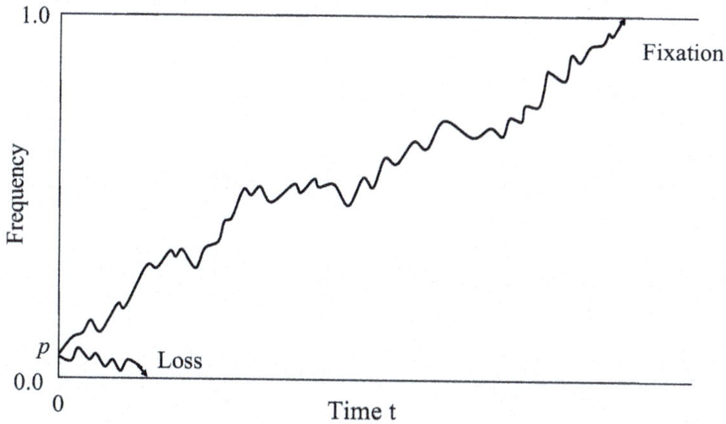

Fig. 6.3 Processes of the fixation and loss of a mutant gene in a finite population

Fixation Probability The case in which this mutation is selectively neutral has been studied in most detail, but in general a single mutant gene that has appeared in the population will, except when natural selection is especially strong, be lost from the population by chance in the majority of cases within about ten generations, whether or not there is selection. Of course, not all are lost, and in lucky cases it survives for many generations and very rarely spreads through the population (becomes fixed) (Fig. 6.3).If the population size is N, the probability of fixation after a sufficiently long time is $1/(2N)$. Thus, in a population of 50,000 individuals ($N = 5 \times 10^4$), the fixation probability is only 1 in 100,000. Moreover, the time required for fixation is known to average $4N_e$ generations (this is the average of only the cases of fixation excluding the cases of loss). Here, N_e is the effective population number mentioned in the previous section, which roughly speaking is the number of breeding individuals in one generation. If the effective number (N_e) is one-half of N, this will take 100,000 generations.

Compared to such a neutral mutation, the probability is much higher that a single mutation advantageous for survival that has appeared in the population will spread through the population, and if this mutant gene has an advantage of s over the preexisting allele, the fixation probability is about twice this, i.e., equal to $2s$ (it is assumed here that $N_e = N$). Therefore, with an advantage of 1% ($s = 0.01$), the fixation probability is 0.02, which is much higher than 1 in 100,000 for the example of neutral case.

In this way, the ultimate fixation probability of a mutant gene differs greatly depending on whether or not there is selection, but there is little difference in its fate during the first few generations. For example for a single mutant gene that is neutral with regard to selection, the probability that it will be lost from the population within seven generations of its appearance is 0.79.On the other hand, if the gene has an 1% advantage in selection over the preexisting one, this probability only changes to 0.78. From this, it can be seen how strongly the behavior of the mutant gene soon after its appearance is determined by chance.

Introduction to Molecular Evolution 7

7.1 Eve of Molecular Evolutionary Studies

Study of Phenotypes The study of biological evolution was, for a long time after Lamarck and Darwin, conducted on visible phenotypes (mainly morphology). The morphologies of extant species were compared; just as important were paleontological studies of the remains of organisms unearthed as fossils; through such studies much valuable knowledge on biological evolution was obtained.

I have already described the course of evolution in Chap. 3; for example, the first ancestor of vertebrates is the Agnatha, which is a lower form of fish that appeared about 500 million years ago around the end of Cambrian. Bony fishes arose from them about 400 million years ago, and one descendent lineage advanced on to land to become the amphibians. The reptiles which are adapted to life on land were eventually born about 300 million years ago. The ancestor of mammals, of which we are a member, was a small organism that emerged about 200 million years ago, and they are imagined to have been similar in morphology and habits to the present day mouse. At the end of Mesozoic, the dinosaurs went extinct and the mammals began their remarkable advance by adaptive radiation. And it is estimated that the human, horse, dog, etc. diverged from a common ancestor a little before this, at about 80 million years ago.

Synthetic Theory of Evolution Darwin's idea of natural selection is even now widely accepted as the basic mechanism for how such morphological and functional evolution occurred. This is as described in detail in Chap. 5. The greatest progress in the theory of evolutionary mechanism after Darwin was made by Mendelian genetics, and it is especially important that the nature of mutations was clarified (see Chap. 4). Eventually, population genetics developed, and the "synthetic theory" of evolution became established in academic circles throughout world. This is, as it were, Darwin's theory clothed in Mendelian genetics. At the beginning of the 1960s when this theory reached its zenith, it gave the impression that the mechanisms of biological evolution were completely solved.

M. Kimura, *My Thoughts on Biological Evolution*, Evolutionary Studies,
https://doi.org/10.1007/978-981-15-6165-8_7

The view that dominated the "synthetic theory of evolution" of that time was an extreme panselectionism, and it was believed that "neutral mutations" which are neither good nor bad under natural selection are virtually nonexistent. Moreover, the view that random fluctuations of gene frequencies, i.e., "genetic drift" explained in the previous chapter, were seldom effective in assembling the genetic composition of a species was dominant. These ideas were based on indirect evidence and panselectionism; there was great uncertainty attached to the method used around that time of deducing genetic change from phenotypic change. Thus, it was impossible to know with certainty the rate at which new mutant genes accumulated in a biological species during the actual course of evolution, or how much hidden genetic variation existed within a population (species).

Neutral Theory However, from around the middle of the 1960s, the methods and concepts of molecular biology (specifically molecular genetics) were introduced into research on evolution and variation. And the situation changed completely. Being able to treat evolution at the molecular level, i.e., at the level of the internal structure of the gene, was epoch-making for evolutionary studies. As a result, various unexpected observations were made, the neutral theory was proposed to explain them, and the position of the panselectionist synthetic theory of evolution, the established theory until then, was forced to be reexamined.

In this chapter, I will briefly explain the facts that are basic to discussing evolutionary mechanisms at the molecular level.

7.2 Basic Knowledge for Understanding Molecular Evolution

DNA Is a Genetic Command The molecular genetic view of life has already been mentioned in Chap. 4, but I would like to briefly summarize the basic facts for understanding molecular evolution and provide additional explanation.

A gene, in molecular terms, is a segment of DNA, occupies a fixed position on a chromosome, and can be regarded as a genetic command written with the four bases, A (adenine), T (thymine), G (guanine), and C (cytosine) as letters, arranged in a row. We can think of a gene as many of these four bases (for example, about 1000) strung together in a straight line. A gene acts in principle as a command for making a protein; three consecutive DNA bases are called a "codon", and each codon specifies one of the 20 kinds of amino acids. However, there are also three codons called "stop codons" that do not correspond to any of the amino acids of a protein, but specify the end of protein synthesis.

When a gene functions, the DNA base sequence of the gene is reproduced as messenger RNA (abbreviated mRNA). In other words, the DNA base sequence of the gene is copied onto the sequence of mRNA. This is also written using four letters, A, U, G, and C, the only difference being that T changes to U. Here U denotes uracil. Roughly speaking, U can be regarded as being the same as T.

Next, this mRNA moves to a cellular particle called the ribosome, an amino acid is attached to each codon, and the amino acids are connected to form a protein. The ribosome is a factory for the production of proteins. The process by which a protein is made according to the mRNA command is called "translation", and the process that comes before this in which the DNA base sequence is copied onto mRNA is called "transcription". Conceptually speaking, transcription only entails changing T to U, but the next step in which this is changed to a sequence of amino acids is complex, and it is appropriate to call this" translation".

The linear arrangement of amino acids is called a "polypeptide", and this is suitably folded to take on a characteristic higher-order structure. A protein molecule formed in this way takes on an important bodily function such as, for example, when hemoglobin carries oxygen to the tissues of the body. The proteins that are made by three dimensional folding become structures in the body, enzymes that promote specific chemical reactions, or substances with other important functions. Here, once the amino acid sequence is determined, the folding is basically automatic and results in the characteristic three dimensional structure (much is still unknown about this mechanism, and it is an important area of research).

In molecular evolution, which is the subject matter of this chapter, the genetic commands change during the evolutionary process. Over many years, the genetic commands characteristic of a biological species written in the four DNA letters undergo changes; at each letter position the DNA bases are substituted, for example what was A becomes T, and what was T becomes C. This takes an extremely long time, and if we look at a particular base site, a change (substitution) eventually occurs after an interval of some several hundred million years. This may be accompanied by the substitution of an amino acid in a protein.

Genetic Code Table I have already noted that three consecutive DNA bases (codon) specify one amino acid; the correspondence relations are given by the genetic code table, and the code table is usually and conventionally expressed in terms of the RNA letters (A, U, G, C). This is Table 7.1. In this table, the first letter of the codon is ordered vertically, the second letter is ordered horizontally, and within each block the third letter is ordered vertically. For example, UUU in the top left hand corner, which is TTT in terms of DNA, corresponds to phenylalanine (Phe). UUU is famous as the first code to be deciphered, which was done by M. Nirenberg. He received the Nobel Prize for his achievement in deciphering the genetic code. UUC which comes next is also a codon for phenylalanine, and UUA below this is a codon for leucine (Leu).

There are 64 (=4 × 4 × 4) places in this table, of which 61 specify 20 kinds of amino acids. If we look at the code for the amino acid valine (Val), we see that, if the first two letters are GU, it corresponds to valine regardless of the third letter.

Degeneracy In this way, a one-to-one correspondence between codons and amino acids is not seen in the genetic code table, and in many cases two or more codons correspond to one amino acid. This is usually called "degeneracy", but in evolutionary discussions two or more different codons that specify the same amino acid are referred to as being mutually "synonymous". In other words, at the DNA level, a

Table 7.1 Genetic code table expressed in RNA characters

1 \ 2	U	C	A	G	2 / 3
U	UUU Phe	UCU Ser	UAU Tyr	UGU Cys	U
U	UUC Phe	UCC Ser	UAC Tyr	UGC Cys	C
U	UUA Leu	UCA Ser	UAA Term.	UGA Term.	A
U	UUG Leu	UCG Ser	UAG Term.	UGG Trp	G
C	CUU Leu	CCU Pro	CAU His	CGU Arg	U
C	CUC Leu	CCC Pro	CAC His	CGC Arg	C
C	CUA Leu	CCA Pro	CAA Gln	CGA Arg	A
C	CUG Leu	CCG Pro	CAG Gln	CGG Arg	G
A	AUU Ile	ACU Thr	AAU Asn	AGU Ser	U
A	AUC Ile	ACC Thr	AAC Asn	AGC Ser	C
A	AUA Ile	ACA Thr	AAA Lys	AGA Arg	A
A	AUG Met	ACG Thr	AAG Lys	AGG Arg	G
G	GCU Val	GCU Ala	GAU Asp	GGU Gly	U
G	GCC Val	GCC Ala	GAC Asp	GGC Gly	C
G	GCA Val	GCA Ala	GAA Glu	GGA Gly	A
G	GCG Val	GCG Ala	GAG Glu	GGG Gly	G

base substitution that results in the same amino acid being retained is a synonymous change.

There are many such examples; the amino acid leucine (Leu) has six kinds of codons, and similarly serine (Ser) has six codons. That is, serine has codons that begin with UC, where the third position does not matter; and in addition there are two codons at a distant place in the code table, which have AG for the first two letters and correspond to serine if the third letter is U or C. Arginine (Arg) also has six codons; four of them are such that if the first two letters are CG the third letter does not matter, and in addition there are two that begin with AG (AGA and AGG).

These three kinds of amino acids each have six codons and hence are called six-fold degenerate. Next, there are five such as valine (Val) and alanine (Ala) that are four-fold degenerate. There are also several that are two-fold degenerate, glutamic acid (Glu) being an example.

Stop Codon and Start Codon Next, a stop codon is a command to terminate reading. That is, amino acids are connected to form a protein, and it instructs the termination of this process. The most commonly used stop codon is UAA, and amino acids can no longer be connected when this is encountered. There are a total of three including UAG and UGA, which are written as "Term." (abbreviation of termination) in the code table of Table 7.1.

The other important codon is the start codon, which instructs initiation of reading, and this is AUG. AUG in the table corresponds to methionine (Met), but when it functions as a codon specifying initiation, it inserts formyl-methionine (fMet) in prokaryotes and methionine in eukaryotes in the production of the polypeptide, at the same time as specifying initiation. When this code AUG is in the middle, methionine is inserted, but when it occurs at the beginning, it becomes an instruction to start. Recently, the DNA base sequences of many genes have been published, and we can see from them that the instructions for making a protein begin with AUG.

Mnemonics for the Code I believe that persons who aspire to do research in molecular evolution should try to memorize this table. Each person can devise his/her own mnemonic method, but one method may be to memorize it like a multiplication table. In particular, for four-fold degenerate codons, it is sufficient to learn the first two letters, and it is possible to memorize them by repeatedly reciting "GG glycine, CC proline, GC alanine ...." In addition, it is desirable to memorize without fail the start codon, AUG, and one of the stop codons, UAA (this author has memorized them by thinking of such phrases as "summer school begins in AUGust", "exclaim oo-ah-ah (UAA) and stop", etc.). Another way that may be useful is to try to remember the code table like a map. The code table usually follows the order U, C, A, G, so it is absolutely necessary to memorize this order. This author recalls this order by reciting "*U*niversal *C*ode *A*ll *G*ood".

7.3 Estimation of the Rate of Molecular Evolution

Comparison of Hemoglobins Research on molecular evolution began with the comparison of homologous proteins between various species. Taking the blood pigment hemoglobin as an example, this molecule exists in the higher vertebrates as a tetramer consisting of two alpha chains and two beta chains, and plays the important role—indispensable for the survival and activity of animals—of carrying oxygen to the tissues.

Of these, the alpha chain in mammals always comprises 141 amino acids strung together. When this is compared between the human and the gorilla, the amino acid sequences are identical except in one place. That is, the only difference is that the 23rd amino acid is aspartic acid in the gorilla whereas it is glutamic acid in the human. Also, when the human is compared to the rhesus monkey they differ in four places; and when compared to the phylogenetically more removed cow, horse, dog, rabbit, etc., 20 more or less amino acid sites differ; but in the other parts the amino acid sequences are in each case completely identical. When we note that there are 20 kinds of amino acids making up a protein, and each site is occupied by one of these, the fact that the hemoglobin alpha chains of mammals are identical at more than 100 of the 141 sites can only be explained if these animals are descended from a common ancestor.

Moreover, differences due to an increase or decrease in the number of amino acid sites (increase or decrease in units of three DNA bases) are first seen when the human and carp, which diverged from a common ancestor almost 400 million years ago in the evolutionary process, are compared; this is a very rare occurrence compared to mere amino acid substitutions. In fact, when the hemoglobin alpha chains of human and carp are compared, differences due to amino acid substitutions account for approximately one half of the total number, i.e., 68, whereas insertions or deletions of sites amount to only three amino acid sites.

Rate of Molecular Evolution Figure 7.1 is a phylogenetic tree of a small number of vertebrates; the geological eras are shown on the left, and the numbers of amino acid differences when the alpha chains are compared are shown below. I will explain with this example how the rate of molecular evolution is obtained from such data.

When the human and horse are compared, the alpha chains of the two differ in 18 places out of the 141 amino acid sites. On the other hand, from paleontological studies, the common ancestor of human and horse is estimated to go back approximately 80 million years. Therefore, the rate at which an amino acid substitution occurs in the evolutionary process is on average

$$\frac{18}{141} \div \left(8 \times 10^{7}\right) \div 2 \approx 0.8 \times 10^{-9}$$

per amino acid site per year. This is roughly 0.8 changes per billion years.

The reason why we divide by 2 in this calculation is because there are two paths from the common ancestor, one leading to the evolution of the human and the other to the evolution of the horse, and the total number of changes in both appear as differences at 18 amino acid sites. However, the above calculation is crude, as it does not correct for the case where an amino acid substitution occurs at the same site along the two paths, or when substitutions occur twice at one site. If the biological organisms that are compared are not phylogenetically too far removed, there will be few amino acid differences in the homologous protein, and the probability of two or more substitutions occurring at the same site is expected to be small. Thus, this calculation method is acceptable. However, in a phylogenetically distant

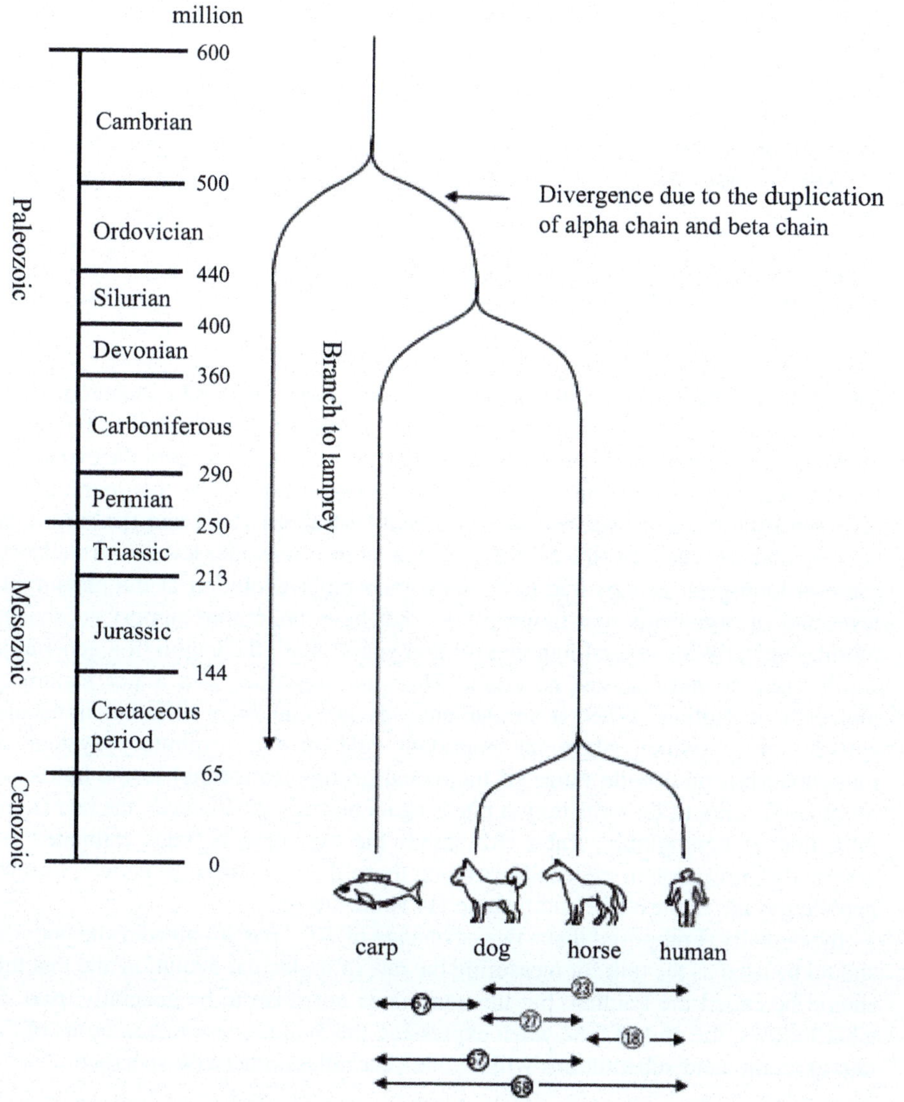

Fig. 7.1 Phylogeny of vertebrates and evolutionary changes in the hemoglobin molecule. The numbers on the arrows show the numbers of amino acid differences when the hemoglobin alpha chain is compared between two species

comparison, for example, when, as in the comparison of the alpha chains of the human and carp, approximately one half of the amino acid sites differ, a correction is by all means necessary. What is important is how to decide whether two or more substitutions have occurred, and various statistical methods have been proposed which I omit here (for details refer, for example, to *Neutral Theory of Molecular Evolution* (Kinokuniya Publisher, 1986, by this author).

Table 7.2 Evolutionary rates of several proteins expressed as the substitution rate per amino acid site per year (from Kimura 1986)

Protein	Substitution rate in units of 10^{-9} per year (unit: Pauling)
Fibrinopeptide	8.3
Pancreatic ribonuclease	2.1
Hemoglobin alpha chain	1.2
Myoglobin	0.9
Insulin	0.4
Cytochrome *c*	0.3
Histone H4	0.01

The Rate of Molecular Evolution Is Constant Using this kind of method, when comparisons are made among the alpha chains, among the beta chains, and between the alpha chains and beta chains of various species of vertebrates, and the evolutionary substitution rates per amino acid site per year are estimated by incorporating the divergence times, all comparisons yield a value of about 10^{-9}; and the surprising conclusion is reached that the rate of evolution at the molecular level is virtually the same in biological groups that have undergone rapid evolution at the phenotypic level and in organisms like living fossils that have undergone almost no change morphologically for several hundred million years. Also, the length of a generation would seem to have almost no effect. This rate constancy is a major feature of molecular evolution. When a similar analysis is done on a different molecule, cytochrome *c* (which plays an important role in the respiratory system of mitochondria), on a wide range of biological groups from fungi to the human, a rough uniformity of the substitution rate is again perceived. However, the rate is one third that of hemoglobin. Table 7.2 shows the evolutionary rates estimated for several different protein molecules. Among those that are currently known, fibrinopeptide has the highest rate and histone H4 is the slowest.

Incidentally, I proposed that a rate of change of 10^{-9} per amino acid site per year should be used as the unit for measuring the rate of molecular evolution and that this should be called the Pauling, but this term does not seem to be generally used. In what follows, the unit of time when expressing the evolutionary rate in terms of the rate of amino acid substitutions will be one year unless otherwise specified.

7.4 Characteristics of Molecular Evolution

Molecular Clock The rate constancy of molecular evolution noted in the previous section applies, not only to the amino acid substitution rates of various proteins, but also to substitution of the DNA bases of genes that produce them. Of course, different genes have different evolutionary rates. The constancy of the molecular evolutionary rate per year is called the "molecular evolutionary clock" or simply "molecular clock", and it has already been mentioned that this is one of the major characteristics of molecular evolution. It is a major advance in biology that, due to this property, it has become possible to construct reliable phylogenetic trees of specific biological groups even when fossil data are lacking.

Rate of Accumulation of Mutations Is Constant That the substitution rates of amino acids and DNA bases in evolution are constant means none other than that mutations at the molecular level accumulate in the species at a constant rate per year during the evolutionary process. I will later explain in detail its population genetic interpretation. Constancy of the evolutionary rate at the molecular level does not hold strictly, and there are some discrepancies and exceptions; nevertheless it is surprisingly constant compared to the evolutionary rate that is observed at the phenotypic level. For example, when the number of mutations causing an amino acid change in the alpha chain of hemoglobin is estimated that have accumulated after the carp and human diverged from a common ancestor (a fish) almost 400 million years ago, approximately the same value is obtained for the evolutionary branch from the common ancestor to the present day carp and for the branch leading to the human. On the other hand, the rates of evolution at the phenotypic level are in striking contrast, and whereas one has retained the structural plan of a fish for nearly 400 million years, the other has achieved a remarkable change in structural plan and evolved to become the human.

In connection with this problem, I predicted in 1969 that the genes of so-called "living fossils" and the genes of organisms that have undergone rapid phenotypic evolution have sustained DNA base substitutions at about the same rate in the evolutionary process at the molecular level, and stated that if this is true it would support the neutral theory. This prediction has subsequently been proven to be correct in many instances. For example, when the alpha chain and beta chain of hemoglobin of the shark, which can be regarded as a living fossil, are compared, the differences are as shown on the left hand side of Table 7.3, and almost the same as the result of a similar comparison of the human alpha chain and beta chain (right hand side of the same table). This shows that after the alpha chain and beta chain arose by gene duplication, most likely about 500 million years ago before the common ancestor of human and shark, mutations accumulated at approximately the same rate in the two biological lineages as they diverged. If the accumulation of mutations were due to positive selection, it would have been influenced by changes of environmental conditions, etc., and there could have been large differences in the evolutionary rates among the various biological lineages, but this is not so.

Conservatism Another major characteristic of molecular evolution is that functionally unimportant (fewer constraints) molecules have a higher rate of evolution. A striking example of this phenomenon is fibrinopeptides A and B. These molecules are released when fibrin is made from fibrinogen in the coagulation of blood, and are thought to have almost no function after they are detached. The evolutionary rate is several times that of hemoglobin, and is the highest value that is currently known for proteins.

An even more interesting example is pro-insulin. This molecule consists of three parts A, C, and B; the part C which occupies roughly the middle third is detached

Table 7.3 Comparisons of the hemoglobin alpha chain and beta chain in the shark and in the human

Type of difference	Shark	Human
0	50	62
1	56	55
2	32	21
3	1	0
Gap	11	9
Total	150	147

The amino acid differences are classified by the minimum numbers—0, 1, 2, or 3—of base substitutions required to explain the change, with reference to the genetic code. Gap indicates the number of amino acid site deletions/insertions.

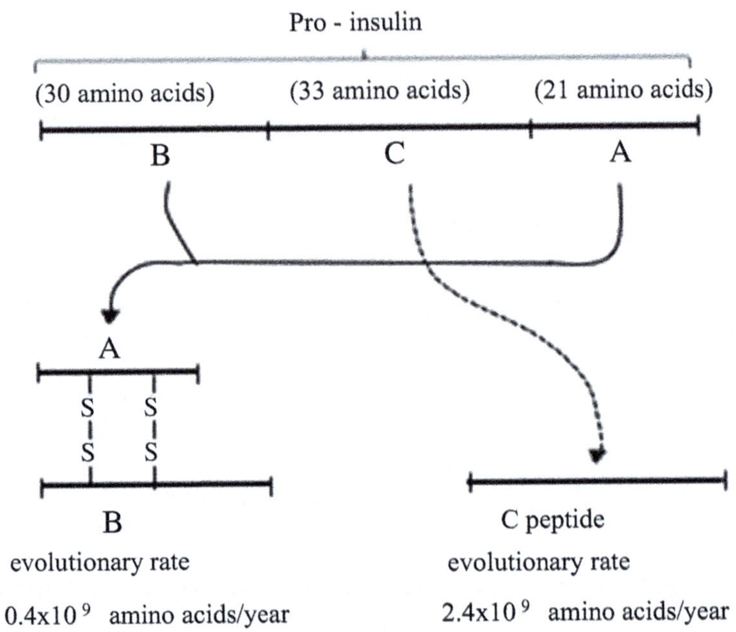

Fig. 7.2 Shows the process by which the middle C segment is removed from proinsulin to produce insulin. The evolutionary rate of the C peptide is several times that of insulin (A-B) (from Kimura 1986)

when insulin is made, and then A and B combine to become the insulin molecule with hormonal activity (Fig. 7.2). For insulin (A-B), it is known that the amino acid substitution rate in evolution is a slow 0.4×10^{-9} per year, but the amino acid substitution rate of part C, which is detached and discarded, is several times the rate for insulin.

As a further example, the alpha chain and beta chain of hemoglobin have similar three dimensional structures, and it is known that the surface part of the molecules is in each case not very important functionally or in maintaining the structure of the molecule. By contrast, the part in the interior called the heme-pocket and its surroundings are the most important parts of this molecule. An investigation of the amino acid substitution rates in evolution shows that, for both the alpha chain and beta chain, the amino acid substitution rate is approximately ten times higher for the amino acid sites on the surface relative to the heme-pocket and its surroundings (when averaged over the alpha chain and beta chain, it is approximately 2×10^{-9} per amino acid site per year on the surface). In general, amino acids are substituted in evolution in such a way that changes in the function of molecules are minimized. This is the property of molecular evolution called "conservatism".

Evolutionary Rate of Synonymous Changes In the above, we dealt with the substitution rates of amino acids in proteins; recently, data on the base sequences of DNA are being published with explosive vigor, and as a result various facts of even more interest are being revealed. One of these is the evolutionary rate of synonymous changes. Here, a synonymous change is as already mentioned a change in a DNA base that does not cause a change in an amino acid, and the greater part of base substitutions at the third position of a codon are of this class.

What is interesting here is the fact that synonymous base substitutions occur very frequently in the evolutionary process. Proteins play a fundamental role in the construction of an organism and in the maintenance of life; if we consider that this function depends on three dimensional structure, which is ultimately determined by the amino acid sequence, a DNA base substitution causing an amino acid change is expected to have a much larger effect in general on the phenotype than one that does not. On the other hand, natural selection acts on the phenotype of an individual and depends on the survival and reproduction of the individual; therefore, synonymous mutations that do not effect a change in amino acids are naturally less likely to be subject to natural selection.

In fact, in the actual evolutionary process, those that do not cause a change in an amino acid have been substituted at a much higher rate than those that cause a change, and have accumulated in the species much more rapidly. It has been shown that proteins such as histone H4 and tubulin (which form microtubules inside the cell) are extremely conservative—the amino acid sequence has changed by about two among 100 amino acid sites in about one billion years—but that at the DNA base level, synonymous changes at the third positions of codons are occurring rapidly. Moreover, the evolutionary rate of these synonymous changes is almost the same as for hemoglobin and growth hormone, which have much higher evolutionary rates at the amino acid level. In other words, it has become clear that the substitution rate in evolution of synonymous DNA bases is not only high, but has the remarkable property that the values are similar for various protein molecules.

To give actual data, the base substitution rates at the third positions of codons of genes such as growth hormone precursor (presomatotropin, on which an enzyme acts to produce growth hormone), hemoglobin alpha chain, histone H4, etc. all lie within an narrow range of $3 \times 10^{-9} - 4 \times 10^{-9}$ per year. On the other hand, there is a large difference in the amino acid substitution rates of these three kinds of proteins, and the rate for histone H4 is less than 1/200 that of growth hormone precursor, which has a high evolutionary rate. In addition, for the gene of the alpha chain of hemoglobin, which has a standard evolutionary rate (amino acid substitution rate), the DNA base substitution rate at the third positions of codons is two or three times that at the first and second positions.

7.5 Accumulation Process of Mutations within a Species

Natural Selection or Genetic Drift What are the mechanisms that produce the substitutions of amino acids and DNA bases in the evolutionary process explained in the previous section? What must be noted here is that such substitutions are not simply the result of mutations at the individual level, but rather the consequence of mutant genes arising at first in a single individual in each biological lineage, then eventually spreading through the entire population (within the species) and being fixed.

From the standpoint of traditional Neo-Darwinism (or the synthetic theory of evolution), the spread of such mutant genes in the population is obviously due to the force of Darwinian natural selection. In other words, if the mutations improve the viability or fertility of individuals that carry them, they spread through the entire population with the help of natural selection. On the other hand, the neutral theory of molecular evolution proposes that the majority of substitutions of amino acids and DNA bases that have been detected for the first time at the molecular level are the result of random fixation by genetic drift of mutant genes that are neither beneficial or harmful under natural selection (selectively neutral).

As a basis for examining which of these two assertions is closer to the truth, let us consider from the standpoint of population genetics the population dynamics of the fixation of mutations at the molecular level, i.e., the problem of how mutants fix and accumulate in the species with the passage of time. What is important here is to clearly distinguish between the appearance of a mutant (for example, a gene with a base change) at the individual level and its fixation (frequency reaches 100%) at the population level. A gene consists of a linear sequence of many nucleotide sites, and molecular evolution deals mainly with mutations due to changes of DNA bases at such sites. In what follows, I will consider the processes by which such mutants are fixed in the population, but a detailed exposition of the theoretical structure of population genetics at the molecular level is of course beyond the scope of this book and will be omitted; I would like to briefly mention only the important results that have been obtained.

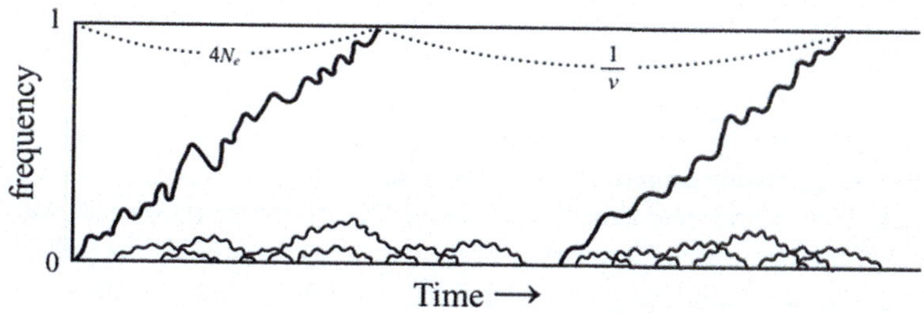

Fig. 7.3 A schematic showing the behavior of a mutation that has appeared in a finite population. The frequency changes of a gene that fixes in the population are shown by thick lines. For fixation of a selectively neutral mutation by genetic drift, the average time from appearance to fixation is $4N_e$ generations, and the interval between two successive fixations is $1/v$ generations. Here N_e is the effective population number, and v is the mutation rate per generation

A convenient mathematical model for dealing with this kind of problem is the "infinite site model" proposed by this author. This is based on the conception that each gene comprises a very large number of nucleotide sites, and assumes that whenever a mutation occurs it occurs at a different site.

Behavior of a Mutant Figure 7.3 shows the behavior of mutants appearing in the population: the majority, including those with a slight advantage under natural selection, are lost from the population within a few generations (for example, ten generations). And the small minority that are lucky spread through the entire population over a very long time (for example, one million generations). As already mentioned in the previous chapter, when a selectively neutral mutation fixes, it takes an average of $4N_e$ generations from its appearance to fixation. Here, N_e is the effective number of the population, which roughly speaking can be thought of as all breeding individuals in a generation. Thus, if we consider a mammalian species with 250,000 breeding individuals, the average time until fixation will be one million generations. In this figure, the frequency change of a mutant going to fixation is shown with a thick line.

Moreover, it can be proven for a neutral mutation that if the mutation rate per genetic locus per generation is v, the average interval between one substitution and the next in the process whereby alleles are substituted in the species is $1/v$. If $v = 10^{-7}$ (one in ten million), the interval between successive substitutions will be ten million generations.

The Equation for the Molecular Evolutionary Rate Let the rate of substitution of mutations in a species, expressed per generation, be k. This expresses the rate at which, with the passage of time, new mutants appear one after another in the population and are fixed; since each substitution takes a long time, we consider a much longer time, and this is, as it were, the average of the number of substitutions (new fixations) per unit time occurring during this time. Therefore, the substitution rate defined in this way is unrelated to the rate at which the frequency of each mutant increases or decreases in the population. What is relevant is the interval between successive fixations, which can be readily understood if we consider the process by which mutants are substituted one after another.

Now, if we consider one genetic locus, and let v be the mutation rate and u be the probability that one mutant appearing in the population at this locus will ultimately be fixed in the population (fixation probability), the following formula

$$k = 2Nvu \tag{7.1}$$

holds. Here N is the actual population number, which is usually larger than the "effective number" (N_e). This formula is based on the following considerations. Each individual in the population carries two alleles descended from the two parents; and there are $2N$ genes at one locus in the entire population, so that $2Nv$ new mutations appear in the population every generation. Noting that the proportion u of these are ultimately fixed in the population, we reason that molecular evolution proceeds at the rate of $2Nvu$. Since we have the infinite site model in mind, we assume that these mutations are all changes at different base sites.

In this formula, the fixation probability (u) depends, not only on the selective advantage (fitness) and degree of dominance of the mutant, but also on population number (N and N_e), and is in general a complex expression. Hence, I will consider in the following two simple but important contrasting cases.

Selectively Neutral Case First is the case of selective neutrality, where the fixation probability is $u = 1/(2N)$, as stated in the previous chapter. This should be apparent if we consider that the $2N$ genes in the population are all equivalent and that it is determined entirely by chance which of these the descendent genes that spread through the entire population are derived from. When we substitute this into formula (7.1), we obtain the very simple formula

$$k = v. \tag{7.2}$$

In words, for a neutral mutation, the substitution rate per generation of a mutant in the population in evolution (evolutionary rate) is equal to the mutation rate per gamete per generation and is independent of population number (it should be noted here that the left hand site of this formula is a quantity related to events at the population level, whereas the right hand site is a quantity related to events at the individual level. Of course, both must be expressed using the same unit of time).

Selectively Advantageous Case On the other hand, as the second case, let us assume that the mutants that accumulate in the species during the evolutionary process are all selectively advantageous, and that they are substituted in the population by the force of Darwinian natural selection. In this case, the fixation probability is $u = 2sN_e/N$,[1] and when we substitute this into formula (7.1), we obtain

$$k = 4N_e sv. \tag{7.3}$$

That is, in this case, the evolutionary rate expressed as the substitution rate of mutations is determined as the product of population "effective number", the advantage under selection of the mutant, and also the rate at which such mutants arise. Thus, it is much more complicated than in the neutral case.

[1]Assume that the actual population number is N; "the effective number" is N_e; and the mutant gene A', appearing in the population, has a selective advantage over the preexisting wild type gene A of s in the heterozygous state (AA') and of $2s$ in the homozygous state ($A'A'$), compared to the wild type homozygote (AA). This corresponds to the case of genic selection described in Chap. 5 (s is the selection coefficient expressing the advantage of A'). If we assume that a single copy of A' initially exists in the population (carried by one individual in the heterozygous state), the probability that this mutant gene will ultimately spread through the entire population and be fixed is given by

$$u = \frac{1 - e^{-2N_e s/N}}{1 - e^{-4N_e s}}$$

(for details, refer to Crow and Kimura, 1970, p. 425, cited at the end of this book). In this formula, in the limit as $s \to 0$, we obtain $u = 1/(2N)$; and on the other hand, in the case of a mutation that has a definite advantage under natural selection and that satisfies $0 < s << 1$, $4N_e s >> 1$, we have approximately $u = 2sN_e/N$. Hence, if $N_e = N$, then $u = 2s$; in other words, J. B. S. Haldane's result mentioned in Chapter is obtained. In addition, the fixation probability of a deleterious mutation that is disadvantageous for survival ($s < 0$) can also be obtained from this formula. In this case, unless the population is small and the detriment is slight (specifically, unless the absolute value of $N_e s$ is of an order not exceeding 1), the fixation probability is an exceedingly small value and in effect can be regarded as almost zero.

8 The Neutral Theory and Molecular Evolution

8.1 Explanation by the Neutral Theory

Explanation of Rate Constancy I have already mentioned that there are two major features of molecular evolution, namely "rate constancy" per year and "conservatism" of the modes of change; how can these features be explained by the neutral theory?

As shown in the previous chapter, for neutral mutations, $k = v$ given by formula (7.2), that is, a simple rule of "evolutionary rate = mutation rate", holds. Hence, the constancy of the evolutionary rate per year can be explained by assuming that the rate of appearance of neutral mutations for a specific gene (or the protein it produces) is constant per year in the various biological lineages. Therefore, if the neutral theory is correct, the rate at which mutations at the molecular level (changes of DNA bases), in particular neutral mutations, occur should be virtually unaffected by environmental conditions, population number, or generation length.

Is the Unit the Generation or Physical Time? The question that immediately arises is the relation to generation. According to the results of research on gene mutations so far, the spontaneous mutation rate is approximately the same per generation—and not per year—among biological organisms such as the human, mouse, and fruit fly, for which the generation lengths differ strikingly. Hence, it is reasonable to expect a large difference in the mutation rates between organisms with short and long generation times when expressed in units of physical time (years). For example, there is at least a 40-fold difference in the generation lengths of the mouse and human, and if the mutation rate per generation is equal in the two species, we can expect a difference of this magnitude in the neutral mutation rates per year. Therefore, when measured with the year as the unit, we might expect that amino acid substitutions would have occurred at a minimum of 40 times the rate in the evolutionary branch leading to the mouse from the common ancestor of the two species than in the evolutionary branch leading to the human. However, this is not the case, and the rates are approximately the same (if we look at the amino acid

M. Kimura, *My Thoughts on Biological Evolution*, Evolutionary Studies,
https://doi.org/10.1007/978-981-15-6165-8_8

substitution rates obtained from actual data, the substitution rate per year in the mouse branch is at most 50% higher than in the human, and the difference if it exists is relatively small).

The fact that the extant results of mutation research and the explanation on the neutral theory are in this way in apparent contradiction has often been used to criticize the neutral theory. Existing data on mutations shows for example that the rate at which a mutation called *dilute* occurs, which causes the color of fur to turn pale in the mouse, and of *albino* in the human are both about 3/100,000 per generation. Moreover, for the spontaneous mutation rates to recessive lethal genes, comparable estimates have been obtained for the human, fruit fly, and mouse when expressed per generation.

On the other hand, in terms of physical time, one human generation is roughly speaking 1000 times longer than that of the fruit fly; and if naturally occurring recessive lethal mutations were to arise in proportion to physical time, they would occur in the human at 1000 times the mutation rate in the fruit fly ($v = 10^{-5}$, see Chap. 6), in other words, at the abnormally high rate of 1% per gene every generation, which does not agree with observed fact. Not only that, but if recessive lethal genes occurred every generation at such a high rate in the human, the total mutational load would be enormous, and the human would be quite unable to persist as a species.

Such considerations suggest that, if the neutral theory is correct, mutations at the molecular level that result in substitutions of DNA bases differ from visible mutations with distinct phenotypic effects and lethal genes, which have been studied so far, in that they occur in approximate proportion to physical time. Whether or not this interpretation is correct must await future experimental research (this is a question that I would by all means like experimentalists to investigate).

Mutation Rate Is Proportional to the Number of Divisions of Germ Cells Recently, estimates are being obtained for the evolutionary rate at the DNA level in the mouse/rat and other rodents with short generation times that are two to three times as high as in the human which has a long generation time. If the neutral theory is correct, the DNA base mutation rate at the molecular level per unit time in the mouse/rat should be two to three times that in the human. This means namely that a "generation effect" exists albeit small.

In this connection, it can be reasoned that the principal cause of such mutations at the molecular level is DNA replication error during cell divisions in the germ line leading to the formation of germ cells. If this is so, such a mutation rate would be proportional to the number of divisions leading to germ cells per unit time, and there should be no problem if it does not differ greatly between the human and the mouse/rat, and if it is not directly proportional to generation length. Recently, Dr. Takashi Miyata of Kyushu University and his colleagues have, based on this idea, investigated whether a difference exists in the synonymous base substitution rates during evolution of genes on the sex chromosomes (X and Y chromosomes) compared to genes on the autosomes, in various biological lineages including the human and mouse/rat; and if such a difference exists, whether this difference agrees with what is predicted from the difference in the number of divisions of germ cells in the female and male; and have obtained results that closely follow expectation.

What is particularly interesting is the evolutionary rate related to the human argininosuccinate synthetase (abbreviated AS) gene. This gene has several pseudogenes (nonfunctional "dead gene", see page 29) produced by reverse transcription, one of which is situated on the Y chromosome (let us call this ψ-Y) and another situated on autosome number 7 (let us write this ψ-7). Both of these genes are descended from an ancestral AS pseudo gene (let this be ψ-a) in the distant past of evolution leading to the human. On the other hand, the normal AS gene is on chromosome number 9, and from a comparison of the base sequences of this gene, ψ-Y, and ψ-7, the result has been obtained that the DNA base substitution rate during the process of evolution on the Y chromosome (ψ-a → ψ-Y) is approximately 2.2 times the substitution rate on the autosome (ψ-a → ψ-7).

Needless to say, the Y chromosome is transmitted only through males. On the other hand, an autosome is on average transmitted once every two times through males. Considering that there are many more divisions of germ cells in males than in females, it is expected that many more mutations due to errors of DNA replication will occur in the germ cells of males. Therefore, genes on the Y chromosome which are transmitted only through males should have twice the mutation rate at the molecular level as autosomal genes which pass through males only once every two times, and if the neutral theory is correct should have approximately twice the evolutionary rate, which agrees with the observed result noted above. The contention of the neutral theory that the DNA base substitution rate during evolution should be directly proportional to the mutation rate is gradually gaining support from many examples.

Explanation by Natural Selection Is Difficult By contrast, in order to explain the constancy per year of the molecular evolutionary rate by Darwinian natural selection, it must be assumed that $N_e sv/g$, which is the product of $N_e sv$ on the right hand side of formula (7.3) divided by generation length (g years) and expressed per year, remains constant in the various biological lineages. The advantage under natural selection of a mutation should depend greatly on the environment in which a species is placed, and it is difficult to understand why the value of $N_e sv/g$ should be approximately the same at the molecular level, among organisms that are extremely different in their phenotypic evolutionary rates. In particular, if a stable environment persists for a long time, it seems natural that a genetic organization that is advantageous in that environment will gradually be achieved (the degree of adaptation will increase), so that the possibility of new mutations arising that are even more advantageous under natural selection than the preexisting alleles should gradually decrease with time. In addition, it is strange that although the population number (N_e) differs among various organisms, its effect is not seen. In this regard, the explanation of the neutral theory is clear and simple.

Mutations that Are Constant Per Generation How then should we understand the observational fact that for visible mutations detected at the phenotypic level and for recessive lethal mutations which have so far been studied, the rate of occurrence among different organisms is approximately constant per generation and not per year?

For these kinds of mutations, it is highly probable that the insertion of a movable genetic element into a genetic locus is an important cause, as mentioned at the end of Chap. 4. Their rate of occurrence perhaps follows rules that differ from base changes at the molecular level (this is also an important research topic for the future).

Explanation of Conservativeness Next, how can the "conservativeness" of change, which is the second feature of molecular evolution, be explained by the neutral theory? To see this, it is convenient to rewrite formula (7.2), i.e., $k = v$, which holds for neutral mutations as

$$k = f_0 v_T. \tag{8.1}$$

Here, v_T is the total mutation rate, of which it is assumed the fraction f_0 are selectively neutral, and the remaining fraction $1 - f_0$ are deleterious and do not contribute to molecular evolution. In other words, $f_0 v_T$ is the neutral mutation rate (v in formula (7.2) of the previous chapter is the neutral mutation rate, and if we denote this by v_0, we can write $v_0 = f_0 v_T$).

Here, I would like to note two points. First, the neutral theory does not claim that all mutations are neutral under selection. The occurrence of deleterious mutations in addition to the neutral is an important premise of the neutral theory. Second, with regard to mutations that are advantageous under Darwinian selection, the existence of such is by no means denied; but the neutral theory assumes that advantageous mutations occur only rarely and can therefore be ignored in considerations of the usual molecular evolutionary rate. Hence, the neutral theory distinguishes mutations that are neutral under selection (i.e., neither advantageous nor disadvantageous) and mutations that are disadvantageous (deleterious) under selection. Of course, we do not mean by a neutral mutation that it is never subject to selection; writing the selection coefficient as s, if the absolute value of s is much smaller than $1/(2N_e)$ such that it is only slightly affected by selection, it can be regarded as "neutral". The reason for this is that the behavior of such a mutation in the population is essentially the same as for a purely neutral one, and its fate is determined almost entirely by random drift.

As to the explanation of the conservativeness of molecular evolution, we can think of this in the following way. The neutral theory assumes only two kinds of mutations, neutral and deleterious; if we regard the neutral ones as "harmless", a functionally more important molecule is subject to greater "constraint", so that a mutation occurring there is more likely to be harmful. Thus, the fraction f_0 of mutations that are neutral decreases as the importance increases.

It should be noted here that, by a neutral mutant gene, we mean a gene that is functionally equivalent to the preexisting one, and that the change makes no difference to survival or reproduction. We are by no means referring only to changes in molecules without function; even for functionally important molecules, a change that substitutes satisfactorily for the preexisting one can be called neutral. Just as in human society it is more difficult to find a replacement for a person in an important occupation, it is natural that for important molecules fewer mutations can be equally suitable.

On the other hand, for a molecule of low functional importance, mutations that impair functions for the survival and reproduction of an individual will occur at a low rate, so f_0 can be regarded as being large. Here, f_0 is a rate (proportion), which can never exceed 1 ($f_0 \leq 1$), and hence, if the neutral theory is correct, an upper limit exists to the evolutionary rate that is determined by the total mutation rate, and the relation

$$k \leq v_T \tag{8.2}$$

should hold. As can be seen from this formula, in the situation where there are no constraints $f_0 = 1$ applies, so that the evolutionary rate should equal the total mutation rate. For synonymous (not causing a change in an amino acid) base substitutions mentioned above, f_0 is fairly close to, but is still <1.

Evolutionary Rate of a Pseudogene In this connection, what is most interesting is the evolutionary rate of the pseudogene mentioned in Chap. 2. To repeat, a pseudogene is a gene that is clearly homologous to a normal gene in its DNA base sequence, but which for many reasons can be regarded as having lost its function as a gene, and is called a "dead gene". Beginning with the discovery several years ago of a hemoglobin pseudogene in the mouse, various examples have subsequently been found.

Studies of the evolutionary rates of hemoglobin pseudogenes have revealed that the evolutionary rates of base substitutions became abnormally high after these genes lost their functions, showing rates that are greater than, and almost twice that of, synonymous substitutions in the hemoglobin genes. Moreover, the rate acceleration is especially marked in the first and second positions (sites) of the codons, and the change is not as pronounced at the third position, which is high to begin with. This can readily be understood if we consider that a pseudogene has no phenotypic effect, that whatever mutations occur there are not subject to natural selection and are all selectively neutral, and that evolution is proceeding at the maximum rate of $k = v_T$.

In 1977 I stated that if the relation in formula (8.2) could be shown to hold for actual data, it would support the neutral theory; it is fortunate that the truth of this claim has been substantiated by the unexpected discovery of the pseudogene and its abnormally high evolutionary rate. The observation that the evolutionary rate at the molecular level is higher when natural selection is less likely to be acting would seem to be explainable only by the neutral theory.

8.2 Intraspecific Variation at the Molecular Level

Polymorphic Phenomena From the standpoint of the neutral theory, intraspecific variation at the molecular level is one phase of molecular evolution, and hence it is necessary to address the question of intraspecific variation. Generally speaking, individuals belonging to the same species are often nearly identical at the phenotypic level. However, it was not known for a long time how much variation existed among conspecific individuals at the level of the internal structure of genes. Moreover, looking at a specific genetic locus, there was no way of knowing whether the normal "wild type gene" was truly a unitary entity, or whether it was actually a collection of many "synonymous alleles" which could not be distinguished at the phenotypic level.

However, from the middle of the 1960s, variation of proteins and in particular enzymes began to be studied by electrophoresis, and it became clear that unexpectedly large molecular variation existed in the fruit fly and human. Subsequently, many studies were conducted using electrophoresis, and in the human an investigation of 71 genetic loci yielded the result that the average heterozygote frequency per genetic locus was 6.7% and that polymorphic phenomena could be seen at 28% of the loci (an enzyme polymorphism is usually called "isozyme polymorphism"). Here "polymorphism" means that two or more alleles coexist at high frequencies in one population. However, it is not called a polymorphism when the allele with the highest frequency exceeds 99% in the population and the frequencies of the remainder total <1%.

A minimum estimate of the total number of genetic loci in the human is about 30,000; so proportionally speaking, the total number of heterozygous genetic loci in an individual should be at least 2000. Variation detectable by electrophoresis does not include all amino acid substitutions but only approximately one third of them. Furthermore, considering that there are also synonymous base substitutions which do not cause an amino acid change, we can infer that at least one million among the approximately 3 billion base sites that constitute the human genome are in the heterozygous state per individual. Thus, when we take two unrelated Japanese individuals, there will be a difference of this magnitude between the base sequences of the genomes of these two individuals.

Extensive studies have been conducted on various species of fruit flies, and average proportions of heterozygous loci between approximately 7% and 18% have been variously reported. It is interesting that the average proportion of heterozygous loci in the many biological species that have been investigated so far fall in the range between 0% and approximately 30%. At one time when data on heterozygote frequency obtained by electrophoresis were limited to the human, mouse/rat, horseshoe crab and a few species of fruit flies, it was erroneously claimed that the average heterozygote frequency lay in the range between 5.6% and 18.4% in almost all organisms. However, it is now known that the average heterozygote frequency in large mammals is generally low, and observed values of 0% are reported for the polar bear, cheetah, and elephant seal.

Controversy Between the Neutral Theory and Balancing Selection Theory The neutral theory views the polymorphic variation detected for the first time at the molecular level as being neutral or nearly so under natural selection and as being maintained in the population in the balance between creation by mutation and random loss by genetic drift. According to the neutral theory, these genetic polymorphisms are merely a transient phase of molecular evolution.

Against this, the "selectionists" strongly opposed the view of the neutral theory from the hitherto mainstream panselectionist stance, and an intense controversy ensued. The selectionists claim that polymorphic variation is actively maintained in the population by balancing selection and is not the result simply of mutation and drift. Here, balancing selection refers to selection that acts in opposing directions, and overdominance is the most important example. Overdominance is the case in which the heterozygote has an advantage under natural selection (higher fitness) over both homozygotes, and, before the neutral theory came to be widely recognized, it received enthusiastic support from many population geneticists as the mechanism for the maintenance of genetic variation.

However, this enthusiasm is now completely gone. As a form of balancing selection that some researchers strongly claim as being important in its place, there is frequency-dependent selection. This was mentioned in Chap. 5; generally speaking, it denotes the case in which the fitness of a genotype is a function of its frequency. However, what is at issue here is something more specific, namely selection in which when the frequency of one of the alleles is reduced for whatever reason, it gains a selective advantage because of this, and as a result neither allele will be lost from the population (for example, when a kind of division of labor exists between the alleles, a situation similar to that in human societies such that life is easy when there are few people with the same occupation, but unemployment occurs when this number increases).

At one time, experimental results supporting the existence of frequency-dependent selection at many enzyme polymorphic genetic loci were variously reported. However, when large scale and careful experiments and observations were later conducted by Professor Terumi Mukai of Kyushu University and his collaborators on polymorphic isozyme alleles of the fruit fly, evidence for the action of balancing selection was not obtained. Rather, much data supporting the neutral theory was obtained. In addition, it became clear from the statistical genetic studies and the analyses of observed data based on them, by Professor Masatoshi Nei and his group at the University of Texas, that the distribution of alleles within species and the distribution of the genetic differences between species could be quantitatively explained from the standpoint of neutral theory.

Nei's Distance Furthermore, Dr. Nei devised a convenient measure expressing the genetic distance between populations of closely related species based on a comparison of isozyme gene frequencies in those populations. This measure is now known as "Nei's distance", and is widely used in the analysis of isozyme polymorphisms.

Dr. Nei computed the genetic distances for Caucasoids, Negroids, and Mongoloids, using data on many protein and blood type genetic loci and obtained

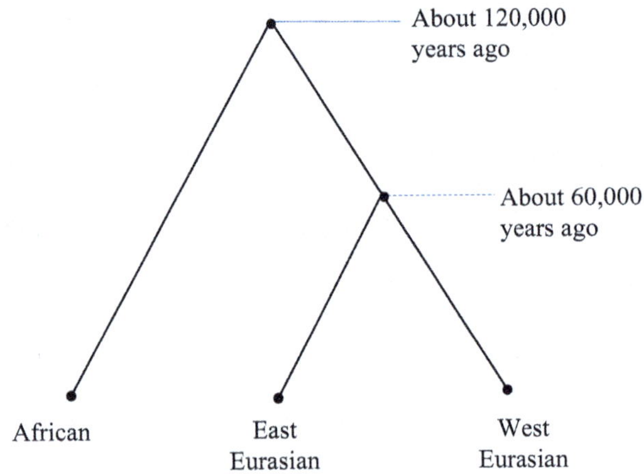

Fig. 8.1 Racial phylogeny based on genetic distance (from a study by Professor Masatoshi Nei)

the following interesting result on the evolution of the races. That is, the Caucasoid and Mongoloid groups diverged from the Negroid groups about 120,000 years ago, after which the Mongoloids and Caucasoids diverged about 60,000 years ago (in other words, Mongoloids are much closer to Caucasoids than to Negroids. See Fig. 8.1).

As an example of similar research, I would like to mention the study by Professor Keiichi Omoto of the University of Tokyo on a race of extremely short people in the Philippines called the Negritos. Professor Omoto examined the blood protein polymorphisms of the Negritos using samples that he collected, calculated the genetic distances from the data obtained, and showed that the Negritos were not, as previously thought, closely related to the Pygmies of Africa, but rather were a race of Asian origin.

Data on DNA Polymorphisms that Support the Neutral Theory Recently, studies of individual variation within a population at the level of DNA base sequences have begun. Of course, it is a very laborious task to actually determine the DNA base sequences of a specific gene or of the chromosomal region that contains it for many individuals in the population, and to date only a very limited amount of data have been obtained. Nevertheless, from the results on DNA polymorphism obtained so far, it can be seen that the parts where the evolutionary rate is high (for example, the third position of codons) also show high rates of polymorphism, which agrees quite well with the prediction of the neutral theory.

Moreover, with mitochondria (abbreviated mt), which are cellular organelles that are involved in the respiratory reactions to produce energy in the cells of higher organisms, it is much easier to extract DNA and to detect individual variation with the help of restriction enzymes than with the DNA of specific genes in the nucleus (a restriction enzyme acts as a "scissor" that cuts DNA at a specific site; for example EcoR1 recognizes the six base sequence GAATTC, and cuts between G and A). For this reason, studies on DNA base sequences of the mitochondria of the human and fruit fly are recently being actively pursued. For example, for the human, A. C. Wilson's group at the University of California has, based on a study of mitochondrial DNA, announced that the ancestor of the present day humans can be traced back to one woman in Africa about 200,000 years ago, attracting much attention (I should add that this claim has been criticized and disputed. Mitochondria are in principle transmitted only from the mother through the egg cells, and are maternally inherited). For *Drosophila melanogaster*, Dr. Sadao Ishiwa's group at Ochanomizu University conducted the first full scale investigation of mtDNA variation in Japanese populations, estimating the amount of polymorphism at the base level (for details see the citation at the end of the book).[1]

8.3 Molecular Evolutionary Clock and Molecular Phylogeny

Molecular Phylogeny The constancy per year of the molecular evolutionary rate is called the "molecular evolutionary clock", and is, as I have repeatedly noted, a remarkable feature of evolution at the molecular level. That we are now able to construct reliable phylogenetic trees using this property, even when fossil data do not exist, is a revolutionary advance in the phylogenetic studies of biological organisms.

The power of such a molecular phylogenetic method was first clearly demonstrated when applied to the problem of when the ancestor of the human

[1]The "infinite allele model" is a model often used to deal with the amount of variation from the standpoint of population genetics, on the assumption that intraspecific variation at the molecular level is neutral with regard to natural selection. It assumes that an infinite number of multiple alleles may exist at one genetic locus, and that whenever a mutation occurs a new allele arises that has never existed before. The motivation for such a model is that, molecularly speaking, a gene comprises many (for example 1000) nucleotide sites, and that, whenever a base changes by mutation, there is a high probability of the change occurring at a different site than before, and that as a result a new allele always arises. Supposing that a gene comprises 1000 nucleotide sites, each of which may be occupied by one of the four kinds of bases, the possible number of alleles is 4^{1000}, i.e., about 10^{602} (in effect an infinite number). Hence, this model applies approximately (this author and J. F. Crow proposed this model in 1964 to deal with variation at the molecular level).

Let us now consider a random mating population, and let its effective number be N_e, and the rate at which neutral mutations arise at a genetic locus per generation be v_0. If we assume in this model that the alleles arising by mutation are all either neutral or deleterious, and that the deleterious ones are removed from the population and do not contribute to variation, it can be proved that the average heterozygote frequency per genetic locus at equilibrium is

$$\overline{H}_e = 4N_e v_0/(4N_e v_0 + 1).$$

When this formula is compared to $k = v_0$, which gives the evolutionary rate, it can be seen that, under the neutral theory, the evolutionary rate and the heterozygote frequency are almost proportional.

Comparing homologous DNA base sequences at a particular genetic locus obtained from several individuals in a population and treating intra-population variation statistically will in future increase in importance. For this purpose a new mathematical theory called the "genealogical process" of a gene is useful. For readers wishing to know details of this, I recommend the review article by Dr. Naoyuki Takahata that is cited at the end of this book.

diverged from the apes, as mentioned at the end of Chap. 3. For readers interested in knowing more, I recommend the book *Origin and Evolution of the Human Viewed from DNA* (Kaimeisha, 1984) by Professor Masami Hasegawa of the Institute of Statistical Mathematics. Professor Hasegawa has recently obtained strong evidence, from the comparison of the base sequences of the eta globin (η-globin) pseudogene, that the ape closest to the human is the chimpanzee.

Phylogeny of 5S rRNA In order to investigate the phylogenetic relations extending over the whole biological world by molecular methods, a molecule like hemoglobin that exists only in higher organisms is inappropriate, and it is necessary to find a molecule that exists in all organisms. One molecule that is ideal for this purpose is ribosomal 5S-RNA (abbreviated 5S rRNA). This is a component of the ribosome, which occurs ubiquitously in the cells of biological organisms, and consists of about 120 bases.

Professor Shozo Osawa and Dr. Hiroshi Hori of Nagoya University (Dr. Hori is currently at Hiroshima University) compared 5S rRNAs in a wide range of organisms extending from bacteria to eukaryotes and obtained many epoch-making results on the phylogeny of organisms. Of course, a phylogenetic tree constructed in this way often agrees with traditional phylogenetic trees based on phenotype or fossil data, but sometimes shows unexpected results. In addition, another major contribution is that it may shed light on unknown phylogenetic relations.

To give one or two noteworthy examples of the results obtained by Osawa and Hori, it was for example shown that methane bacteria and halobacteria, which C. Woese and others regarded as the oldest bacteria and named "archaebacteria", were by no means an old bacterial lineage, but rather were new (close to eukaryotes). Based on this result, Osawa and Hori have proposed that the term archaebacteria be abandoned and that they should be called "metabacteria" instead. As an example in plants, the hitherto established theory that mosses evolved into ferns was not supported by the studies on 5S rRNA, it being more likely that mosses are a degenerate form of ferns.

DNA-DNA Hybridization A method called DNA-DNA hybridization is also extremely useful for molecular phylogenetic studies of a wide range of biological species. An outline of this method is as follows: first, extract DNA from the biological species that are to be compared, make fragments of this DNA, and separate the DNA double strands into single strands by heating them. Next, anneal the single strand DNAs of different species made in this way utilizing their homology; and then after these so-called DNA-DNA hybrids have been formed, estimate the degree of difference in the base sequences of the two chains by measuring their thermal stability.

The American ornithologist C. G. Sibley used a convenient DNA hybrid analyzer for making these measurements, and investigated the phylogenetic relations of 1600 species of birds by comparing more than 20,000 combinations of DNA. One interesting result of this study is that it has become clear that vultures such as the condor, which live in the New World, are phylogenetically different from the hawk, but rather are closely related to the stork.

Evolution of the Influenza Viruses The nuclear DNA of the higher organisms that have been discussed so far has an evolutionary rate that is extremely slow by the standard of our daily lives, and in order to see changes in a gene it is necessary to compare organisms that have diverged from a common ancestor at least several million years ago. On the other hand, recent studies have shown that RNA viruses in which the genetic material is not DNA but RNA evolve at approximately one million times the rate of a DNA gene. A good example is the influenza virus (especially type A) analyzed by Professor Masatoshi Nei and his collaborators at University of Texas and by Dr. Takashi Miyata's group at Kyushu University.

Influenza type A virus is known to cause influenza epidemics. When the number of years between different epidemics and the number of base substitutions that have accumulated during this time in its genes are examined for a correspondence, a clear linear relation is obtained, showing that its evolution also obeys the molecular clock.

Moreover, it has become clear that base substitutions that do not cause an amino acid change, i.e., synonymous substitutions, occur at higher frequencies than non-synonymous substitutions in which amino acids change. Thus, in this case also the typical aspects of neutral evolution are seen. Here "neutral evolution" refers to evolution by the incessant appearance of mutants (also called mutational pressure) and their random fixation in the population by genetic drift. Miyata and others estimate that the synonymous substitution rate of genes of the influenza virus averages approximately 1.1×10^{-2} per base site per year. Since, by contrast, the evolutionary rate of synonymous substitutions for the nuclear genes of higher organisms is generally about 3×10^{-9}~4×10^{-9}, molecular evolution is occurring in the influenza virus at about two million times this rate. On the neutral theory, the reason for this is that the mutation rate is high (i.e., v_T in formula (8.1) is large), which agrees with the fact that the mutation rate is generally very high in RNA viruses. In this connection, RNA viruses are not equipped with a developed mutation repair mechanism as is the case for DNA (see Chap. 4 for mutation repair).

Evolution of the RNA Oncogene and AIDS Virus A similar result has been obtained from a study of the molecular evolution of the oncogene. Dr. Takashi Gojobori of the National Institute of Genetics and Dr. Shozo Yokoyama of the University of Illinois have succeeded in comparing the evolutionary rate of an oncogene (abbreviated v-onc) carried by an RNA cancer virus (retrovirus) and the evolutionary rate of the homologous oncogene (c-onc) on the DNA genome of the host cell, and obtained a result showing that the former was likewise evolving at approximately one million times the rate of the latter. This can also be explained by the neutral theory if we consider that the mutation rate of the RNA virus is about one million times higher than that of nuclear DNA. Furthermore, that synonymous base substitutions are clearly occurring at a higher rate than non-synonymous substitutions agrees well with the neutral theory.

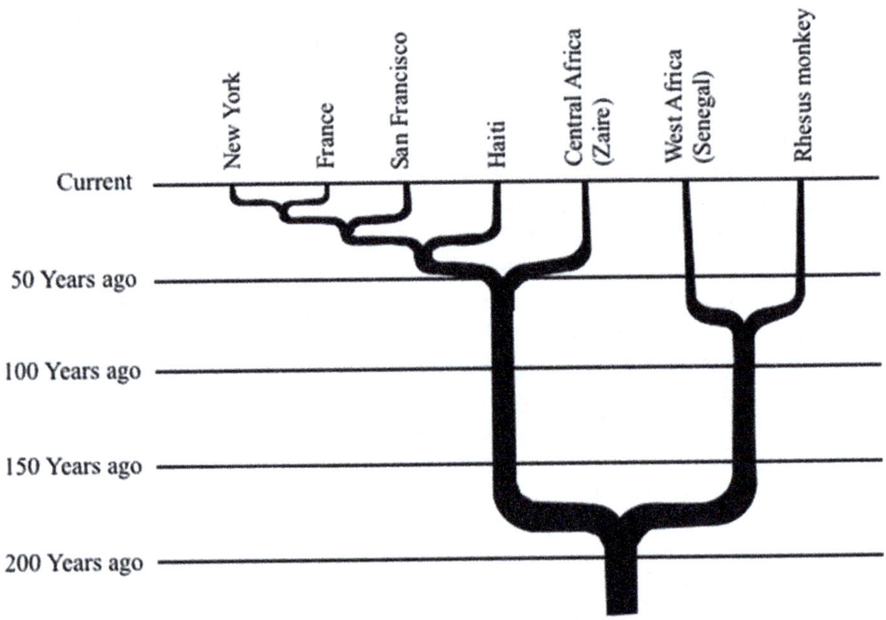

Fig. 8.2 Evolutionary phylogeny of the AIDS virus (from Dr. Takashi Gojobori). The human AIDS virus that initially spread from central Africa form the HIV-1 group (includes Haiti, France, and the United States), whereas the new human AIDS virus and the Rhesus monkey virus form the HIV-2 group. These two groups are thought to have diverged from an ancestral virus more than 150 years ago

AIDS (acquired immune deficiency syndrome), a disease that is recently creating a stir, is caused by the AIDS virus, which is a type of retrovirus with an RNA genome (RNA cancer virus) and which also shows an extremely high evolutionary rate. Dr. Gojobori and his co-workers compared the base sequence data of the genes of the AIDS virus isolated from AIDS patients in various regions of the world, and obtained a molecular phylogenetic tree that includes other closely related retroviruses (see Fig. 8.2). From this study a result was obtained strongly supporting the claim that the human AIDS virus originated in Africa.

8.4 Other Topics Related to Neutral Evolution

Biased Usage of Synonymous Codons In this book I have referred many times to synonymous (not changing amino acids) base substitutions in the evolutionary process, and stated that the fact that their evolutionary rate is much higher than that of non-synonymous (changing amino acids) base substitutions is evidence supportive of the neutral theory. However, it is not the case that no objections were raised against this in the past, and some scholars have invoked the phenomenon of the "biased usage" of synonymous codons to express strong opposition against the explanation of the neutral theory.

First, let me explain with an example what is meant by the biased usage of synonymous codons. There are six kinds of codons that code for the amino acid leucine, and as can be seen from the genetic code Table 7.1, UUA, UUG, CUU, GUC, CUA, and CUG all correspond to leucine. On the other hand, when we examine the frequencies with which these six kinds of synonymous codons are found in the base sequences of various genes in various organisms, the distributions are not uniform (same frequency) and in many cases are biased toward a specific codon. For example, in *E. coli*, CUG is most frequently used among these six kinds to code leucine, but in yeast, UUG is used at maximum frequency. Opponents criticize the neutral theory as being wrong, because, if the neutral theory is correct, these six kinds of synonymous codons should be used at equal frequencies, but in fact are not.

However, fortunately, studies by Dr. Toshimichi Ikemura of Kyoto University (currently National Institute of Genetics) have revealed that the problem of the biased usage of codons can be satisfactory understood within the framework of the neutral theory. Dr. Ikemura measured the relative amounts of various transfer RNAs within a cell in *E. coli* and yeast, and discovered a strong correlation between this quantity and the usage frequency of the corresponding codons. Here transfer RNA (abbreviated tRNA) is an RNA molecule that plays the role of a "transporter" of the necessary amino acids when the base sequence of messenger RNA is translated into the amino acid sequence of a protein.

For each of the 20 kinds of amino acids, there exists one or several tRNAs. In *E. coli*, 26 kinds are known; there exist four kinds of tRNA molecules for the six synonymous codons of leucine, and the tRNA variety among these four kinds corresponding to the codon CUG exists in an overwhelmingly large amount within the cell of *E. coli*. On the other hand, in yeast, it is known that there are three kinds of tRNAs that transport leucine, and of these the tRNA corresponding to UUG exists in the largest amount (the relative amounts of these tRNA molecular varieties are determined by the numbers of genes that make them). Therefore, the biased usage of synonymous codons can be considered as the optimal state in which an appropriate division of labor has been achieved among the various synonymous codons so as to maximize translation efficiency. In other words, it is an adapted state in which the various tRNAs existing in a cell are not left wastefully idle; and once this state is reached, neutral evolution subsequently proceeds so as not to corrupt this state, and base substitutions in genes that lead to large deviations from this state are subject to negative selection (i.e., in $k = f_0 v_T$ given by formula (8.1), f_0 becomes small). Readers wishing to know more about this problem are asked to refer to my book *Neutral Theory of Molecular Evolution* (Kinokuniya Publisher, 1986).

Exceptional Genetic Codes Dr. Ikemura coined the term "dialect" to refer to the biased usage of synonymous codons, in particular the idiosyncratic usage frequencies of various organisms. An example deserving special mention, in which such "dialect" formation has gone even further and a "language" has emerged that is incomprehensible elsewhere, is the exceptional genetic code usage system discovered in *Mycoplasma capricolum* by the group at Nagoya University including Professor Shozo Osawa and Dr. Akira Muto. *Mycoplasma* is a small and singular bacterium that parasitizes eukaryotes; its DNA has an extremely A-T rich base composition, and approximately 75% of the base pairs in the genome are pairs consisting of A and T. G and C pairs constitute only the remaining 25%.

One observation of interest here is the usage frequencies of synonymous codons; the third positions of the codons are almost all A or U, and C and G are rarely used. For example, of the six synonymous codons for leucine, UUA has an overwhelmingly high usage frequency. This situation can be readily understood by analogy with a printing house with an extreme shortage of types corresponding to some of the letters, and which substitutes for types in short supply with other types (provided the meaning is unchanged). What is even more surprising is the fact that one of the stop codons, UGA, which in the standard genetic code of Table 7.1 is supposed to command termination, is used as a code for one of the amino acids, tryptophan (Trp).

For a long time, the genetic code Table 7.1 was believed to apply universally, without exception, to all organisms on the Earth. However, less than 10 years ago, exceptional codon usage was discovered in the DNA of mitochondria, which was big news. But this was in a cellular organelle, which contains genetic information that is of secondary importance to an organism. By contrast, it was in *Mycoplasma* that the existence of exceptional codon usage was first discovered in the main body of DNA that oversees the genetic information of the cell.

Around the same time, researchers in the United States, Germany, and France discovered exceptional codon usage in the nuclear genes of the ciliates, *Paramecium* and *Tetrahymena*; the stop codons UAA and UAG were both read as glutamine (Gln). I would like to add that in Japan also the group at the National Cancer Center including department head, Susumu Nishimura, and Dr. Yoshiyuki Kuchino demonstrated, by detailed structural analysis, the existence of tRNAs that translate these two codons.

Explanation by the Neutral Theory It is an extremely interesting question how the exceptional codon usage of *Mycoplasma* evolved. Does this remarkable phenomenon really represent the vestige of a primitive genetic code?

Unexpectedly, it was pointed out by T. H. Jukes that such a discrepancy could, without contradiction, be explained from the standpoint of the neutral theory as a secondary product of evolution. That is, it is explained as having been derived by neutral evolution from the universal code by biased mutation pressure.

To explain in more detail, in the DNA of *Mycoplasma*, the mutation rate from G and C base pairs to A and T pairs has become overwhelmingly higher than the mutation rate in the reverse direction (this may be due to a preceding evolutionary change in DNA polymerase), and with this as the primary cause, the current anomalous code evolved by a series of neutral changes to be described below. What is crucial is how the preexisting codon UGG for tryptophan came to be almost completely replaced by the stop codon UGA (a small number of UGGs apparently remain and are used).

According to Jukes, in the first step, the stop codon UGA changes to a different stop codon UAA under G→A mutation pressure. That is, the role played by UGA is completely taken over by UAA, and UGA as a code for termination temporarily disappears from the genome (unless this happens, UGA will later be read as tryptophan, and superfluous amino acids will be connected to the protein causing havoc).

In the succeeding second stage, the preexisting tryptophan tRNA gene duplicates, and one of two tRNA genes made in this way changes to recognize the codon UGA and to transport tryptophan. This is due to a change at a part of tRNA called the anticodon, which recognizes the codon (for details, see Jukes (1985), Yamao et al. (1985), and Yamao (1986)). Such a change in the anticodon (CCA→UCA) also occurs by biased mutation pressure (C→U), giving birth to a new tRNA. The determination by Professor Osawa and others, by an analysis of DNA base sequences, that the new and old tRNA genes are located next to each other is a valuable finding.

Given such changes, no harm is done if the orthodox tryptophan codon, UGG, gradually changes to UGA under biased mutation pressure (G→A), because a new tRNA exists to read it as intended as tryptophan. On the other hand, the old tRNA which recognizes UGG loses its job, as it were, and in Jukes' words comes to exist only as a vestigial organ. This series of changes is not deleterious, and can be understood without difficulty as neutral evolution under biased mutation pressure. It is likely that various such examples will emerge in the future.

Models of the Origin of Life If we pursue the evolution of the genetic code further, we will ultimately confront the question of the origin of life. The elucidation of the origin of life is one of the major problems that modern natural science is required to deal with, and due to the development of molecular biology, it is now becoming a problem within our grasp. This is, of course, one of the biggest problems for molecular evolution as well. It would be splendid if in future young Japanese researchers will be able to make world-class contributions in this area.

The German chemist and Nobel Prize winner, M. Eigen, has been vigorously tackling this problem both experimentally and theoretically for more than 10 years. The hypercycle model of self-replicating systems proposed by Eigen is well-known. This is an entirely deterministic treatment assuming Darwinian natural selection (for details, see reference cited at the end of the book), and the role of randomness is not considered.

Dyson's Model On the other hand, the eminent theoretical physicist, F. J. Dyson, of Princeton University has recently combined the classical Oparin theory, Margulis' theory of symbiotic evolution, and my neutral theory, and furthermore used my mathematical methods dealing with the stochastic processes of gene frequencies, to propose a new theory that might be called the "two stage theory of the origin of life". The essential points of Dyson's model are as follows.

First it is assumed, as in Oparin's theory, that a "cell" comprising an inactive droplet exists, inside which many monomers such as amino acids are to be found and combined in various ways to make macromolecules. At each site of the macromolecule, random substitutions of monomeric residues are constantly occurring; each monomeric residue can be classified as being in the "active" state or the "inactive" state, and when an active form is produced by chance, the probability that an active monomeric residue will be substituted at other sites increases. In other words, this is a kind of enzymatic activity; the state in which macromolecules comprising only active monomers are fixed in the cell is called "living", and the opposite state comprising only monomers without activity is called "dead". The cell goes back and forth between living and dead, taking a long time to do so; Dyson obtained the time required to do so (T), using a mathematical method similar to the one used to compute the fixation time by random drift of a mutant in a finite population, and determined the kinds of monomers and the sizes and properties of the macromolecules acting as enzymes that would be required for this process to occur within the time available on the Earth (for details see the citation at the end of the book). This hypothesis proposes that the emergence of such a "living" state was the first stage in the origin of life, and that in the second stage RNA attached itself like a parasitic organism to such a living cell and eventually evolved into genes.

As the well-known experiment by S. L. Miller shows, amino acids are easily formed just by discharging electricity into a mixture of methane, ammonia, hydrogen, steam, etc., which is believed to be the composition of the primitive atmosphere; moreover, whereas amino acids link together with comparative ease, not only are nucleotides, which are the components of nucleic acids, not readily formed, but rarely do they link together without the help of enzymes; hence I think that Dyson's model is reasonable. Dyson's small book entitled *Origins of Life* (1985) describes clearly the molecular biological research to date on this problem and is readable.

Very recently (1986), an example in which an RNA molecule itself has enzymatic activity was discovered by Thomas Cech of the United States and his collaborators, receiving worldwide attention. As a result, it was claimed that the origin of life could be understood with only RNA molecules, without taking into consideration other molecules (in particular proteins); but because the question remains as to how the molecule with the special RNA base sequence possessing enzymatic activity first came into being, it seems to me too simplistic.

8.5 Bridging Molecular Evolution and Phenotypic Evolution

Differences Between Molecular Evolution and Phenotypic Evolution I have already mentioned that a new perspective on research on biological evolution unfolded when it became possible to study evolution at the molecular level. In addition, the molecular biology of eukaryotes has developed explosively during the past few years, and new findings are being obtained with dizzying rapidity. It is urgently necessary to develop the fields of molecular evolution and population genetics incorporating these findings, and I hope that in Japan as well as in Europe and the United States young researchers will rise to the challenge. Another important issue that I feel is particularly important biologically is how to relate evolution at the molecular level to evolution at the phenotypic level.

As already mentioned, conspicuous differences exist between evolution at the molecular level and evolution at the phenotypic level. First is the evolutionary rate, which at the molecular level has a property called the molecular evolutionary clock; in organisms such as "living fossils" that have undergone almost no changes phenotypically during several hundreds of million years and in organisms that have rapidly evolved at phenotypic level, when we look at the same molecule (gene), base substitutions are occurring in both at almost the same rate per year.

The second is that conservativeness of change is a major characteristic of molecular evolution. Namely, molecules have changed so as not to impair the functions or structures of preexisting molecules, according as it were to "don't rock the boat principles".

By contrast, the characteristic of evolution at the phenotypic level is, in one word, "opportunism". For example, if we think about wings in animals, the wings of birds are derived from front limbs, whereas insects have wings of entirely different origin, and the material is of no consequence so long as it is useful for flying. Moreover, so-called "parallel evolution" or "convergent evolution" are manifestations of opportunism. For example, in the Eutherian mammals which have a placenta, numerous species arose by adaptive radiation; and in Australia a radiation of the marsupials occurred, such that dog-like, squirrel-like, and anteater-like, etc., adaptive forms that are surprisingly similar to the Eutherian mammals arose. Such opportunistic evolution can only be explained by invoking Darwin's natural selection theory. That is, this can be regarded as the result of genetic variants that by chance are advantageous in the environment leaving numerous descendants and of the disadvantageous ones perishing, resulting in directional selection as described in Chap. 5.

The problem is how to understand the large differences between the features of molecular evolution and phenotypic evolution; in my opinion, what is most commonly seen at the phenotypic level is stabilizing selection that maintains the status-quo (see Chap. 5); if many genes of small effect are involved in each phenotypic trait, the strength of selection per locus will be extremely small at the equilibrium state in which the optimum and the mean approximately coincide, and there will be sufficient room for neutral evolution to occur at the molecular level. In other words, many genetically different combinations can produce phenotypically identical forms, so that there is a much greater degree of freedom at the molecular level than at the phenotypic level.

Micro-Evolution, Macro-Evolution and Speciation A distinction is often made with regard to evolution at the phenotypic level between "micro-evolution" and "macro-evolution". Micro-evolution refers to a genetic change in the population (species) that can be observed within our lifetimes, and industrial melanism of the moth described in Chap. 2 is a good example. Macro-evolution usually refers to the formation of a new species, or to changes that transcend species, and sometimes vaguely to phenotypically large evolutionary changes. The late Dr. Taku Komai proposed that the intermediate phenomenon of speciation should be distinguished from macro-evolution and called meso-evolution.

A detailed exposition of the question of speciation cannot be made in this book due to lack of space, but we can say that the fundamental factor in the formation of a new species (speciation) is "isolation". It is convenient to classify isolation into geographical isolation and reproductive isolation. First of all, when a population is separated from the source species by a sea, a major river, a mountain range, etc., (geographical isolation), new genetic variants become fixed within that population by random drift or regional adaptation (selection). Next, with this as the trigger, genetic properties (reproductive isolation) such as difficulty of mating with the source species, infertility even when interbreeding occurs, or the lowering of fitness of the hybrid develops, and the distinctiveness as a new species is established. After this, the new species will evolve by accumulating genetic variants peculiar to it one after another. A famous example of speciation by geographical isolation is, as once observed by Darwin, the case of Darwin's finches, which are unique to each of the islands of the Galapagos Archipelago.

Evolution at the Phenotypic Level When we consider evolution at the phenotypic level, what is indisputably interesting is macro-evolution and the associated question of evolution at the phenotypic level. In this case, Darwinian natural selection undoubtedly plays the major role, but the simple panselectionism that was entertained during the golden age of the synthetic theory of evolution needs to be revised.

Partly in reaction to the fact that the great majority of base substitutions in the so-called structural genes that code for proteins, etc., are now considered to be the result of neutral evolution, it is nowadays widely claimed that what is involved in phenotypic evolution are changes in "regulatory genes". Regulatory gene is the blanket term for genes that function to regulate the expression of other genes, and DNA signals that regulate transcription can be included among them (for transcription regulation refer to the review by Fujita et al. cited at the end of the book). This kind of thinking is essentially the same as the view often expressed in the age of classical genetics that modifying genes and quantitative trait polygenes are important in evolution.

In another vein, stimulated by the discovery of homeotic genes involved in development (morphogenesis) (see Chap. 4), it is recently actively argued that "developmental constraint" is important in evolution. This author feels that although such arguments are not meaningless, it should not be forgotten that it is selective constraint that directly relates to evolution (it is a kind of argument that, in relation to admission to a university for example, focuses on the educational environment of the home and forgets the importance of the grades on the entrance examination).

Causes of Macro-Evolution What are the factors that directly contribute to macro-evolution? To give two especially remarkable examples of evolution among the evolutionary pathways in the past described in Chap. 3, one is the emergence of multi-cellular organisms (in particular animals) during the Precambrian period, and the other is the adaptive radiation of mammals which began in the Cenozoic era. In the former case, a new life form, the multi-cellular organism, was pioneered, and the seas were transformed into one large experimental arena for the evolution of higher organisms. In the latter, the various ecological niches occupied by the dinosaurs became vacant due to their extinction, and the surviving mammals radiated explosively into them.

In short, it is the emergence of new environments. When an assemblage of various species of organisms is placed for a long time in a stable environment, its members are squeezed into their respective environments and are so to speak restrained by stabilizing selection. Thus, for a major evolutionary change to occur, it is a precondition that they first be liberated from this situation. I would like to call this "liberation from selective constraint". In addition, it is necessary that a new rich environment without competitors exists (macro-evolution is almost always linked to the exploitation of a new adaptive regime). If no land of appropriate size existed on the Earth, no matter how often gene duplications were repeated in the genomes of organisms in the seas, higher organisms would not have advanced beyond bony fishes. I would like to emphasize this point, because when geneticists argue about evolution, their attention is limited to the structure of the genome and tend to forget the importance of environmental factors (this author is also a researcher of genetics).

I would like to point out one other thing, which is that functionally wasteful DNA such as is seen in higher organisms is rare in bacteria, and that bacteria in this respect seem to be constructed highly efficiently. Here, rapid division and reproduction has become almost the sole criterion for selection, and everything other than what is required to win in this competition has been discarded, as a result of which any scope for other properties to develop has been lost; bacteria are forever unable to progress beyond the primitive unicellular organism (such an example suggests to me that overly severe competition in the entrance examination may be harmful for social progress).

In any case, for large scale evolution to occur, it is necessary that natural selection be temporarily relaxed, which is a point often overlooked by the panselectionists. Recently, some researchers of population genetics have applied formulas for selection on quantitative traits (formula (5.3) of Chap. 5 or an extension thereof to two or more traits) to actual data on evolution, for example the rate of increase of body size in dinosaurs, and estimated various parameters, believing that by doing so they can elucidate the mechanisms of phenotypic evolution. In my view such a formalistic approach is not very fruitful. What is really more important is to clarify under what circumstances selection operates (or does not operate).

Role of Genetic Variation With regard to the appearance of genetic variation which is the material for evolution, the higher organisms possess a vast potential for generating variation. This is quite clear when, for example, we consider the dog, which was formed by domestication of the wolf. Just considering body size, it is surprising that such a wide variety, from the Mastiff which can reach 100 kg to the Chihuahua which is less than 1 kg, have been formed (for the evolution of the dog, see *Dogs* translated by Hiroshi Sawazaki from the book by J. C. McLoughlin, Iwanami Publisher 1984).

Of course, it is one-sided to emphasize only the role of the environment in evolution, and the role of genetic variation must also be elucidated at the molecular level. Recently, many instances of the existence of unexpected relations between molecules thought to be totally unrelated have been revealed from a comparison of DNA base sequences (or the amino acid sequences of proteins). For example, it has been reported that the protein component, crystallin delta, of the lens of the eye and the enzyme arginosuccinate lyase which is involved in the urea cycle have approximately 60% homology in their amino acid sequences.

I am impressed by the unexpected versatility of molecules as seen in this example. In any case, the elucidation of the mechanisms by which a molecule with a new function emerges will hereafter be an important topic for research in molecular evolution and population genetics. In this case also, it is certain that not only gene duplication followed by neutral evolution, but Darwinian natural selection as well, acts; the issue is how it acts. I look forward to developments of future research.

An Evolutionary Genetic World View

9

9.1 The Human as a Product of Evolution

Gene Pool and Cultural Heritage It is often said that each of us is a unique, irreplaceable being, and the biological basis for this claim can, first of all, be sought in the vast genetic variability present in human populations. With the exception of identical twins, it is unthinkable that a person who is totally identical genetically will reappear in human history. If we treat the entire collection of genes possessed by all mankind or by each ethnic group abstractly and call it the "gene pool", each individual is as it were a random sample drawn by chance from this vast gene pool, and with regard to genetic composition it may be said that we are all born "unequal".

Of course, in addition to existing as biological organisms, humans have a developed culture, and the amount of cultural information that has accumulated in human society since the birth of civilization is truly enormous. Furthermore, our heads are full of knowledge obtained after birth through experience and education, which is unrelated to genes, and this is also unique to each individual. Existence of the self is central to the "consciousness" of the individual; but from the standpoint of evolutionary genetics, the entire gene pool and cultural heritage of humans is fundamental and permanent, whereas the individual may be likened to one runner in a long distance relay. However, we need not cite examples such as the geometry of Euclid or the music of Bach to realize that the role of prominent individuals in human cultural evolution is truly great.

Four-Billion-Year Evolutionary Legacy On the other hand, from the purely biological standpoint, humans are carriers of an evolutionary legacy going back 4 billion years since the birth of life. For example, we have two arms and two legs, and this structure has been handed down from a fish with fleshy leaf-like fins called crossopterygians of approximately 370 million years ago (the famous coelacanth is a surviving member of this group). At that time (the Devonian Era), there were many

M. Kimura, *My Thoughts on Biological Evolution*, Evolutionary Studies,
https://doi.org/10.1007/978-981-15-6165-8_9

shallow swamps, and droughts often occurred seasonally; this kind of fish (for example, *Eusthenopteron* has been well studied) was able, with the help of its strong fore and hind pairs of fins, to crawl out of dried up swamps and move to other ponds with plenty of water. It is believed that this was advantageous for survival and reproduction. It is said that the principal skeletal structures of the four limbs of all terrestrial vertebrates that are its descendants, including the present day human, can already be seen in the structure of the fins of the *Eusthenopteron*.

The Himalayan Mountain Range is apparently about 50 million years old, and it is notable that the structure of the four limbs of vertebrates is much older than this. The biological properties that we possess have all been handed down from the distant past, and the culture of humans has been built on this evolutionary heritage.

Development of Intelligence Needless to say, the greatest characteristic of the human is a developed intelligence. During the process of human evolution over the past three million years or so, brain size (more exactly endocranial volume) has increased approximately two-fold, which is an exceptionally high rate of phenotypic evolution. Incidentally, following J. B. S. Haldane, a unit called the Darwin is sometimes used to express the rate of phenotypic evolution; this is the rate at which size increases approximately 2.72 (the value of the base, e, of the natural logarithm) times in one million years. Using this, the evolutionary rate of brain size is 1/5 Darwin. This is ten times faster than the evolutionary increase of the body length of dinosaurs during the Mesozoic era, which averages 26 milli-Darwins. From this we can surmise that intelligence played an extremely important role in the survival and reproduction of hominins in the past. In particular, there must be difficulties when a baby with a big head is born from the mother, and evolution toward a big head would not have occurred unless it was very advantageous under natural selection. In other words, a situation in which superior intelligence conferred a big advantage under natural selection in the Darwinian sense persisted for a long time.

Generally speaking, intelligence is a property that has developed only in animals that move around to catch prey and in search of mates; intelligence did not develop in plants which are unable to move by themselves. Animals have evolved through a struggle of "eating or being eaten". That is, carnivorous animals have evolved to catch prey just a little more skillfully, and prey animals to escape from enemies just a little more skillfully. During this process the brain (intelligence) gradually developed to gather information from the surroundings appropriate for each purpose and as a computer to direct the most advantageous behavior for survival based on this information. In the terminology of computers, the hardware was drastically improved.

When we compare the structures of the cerebra in various animals from fish to the human, we are able to understand clearly how the brain developed in the evolutionary process. Our ability to recognize the external world, as defined by philosophy, temporally and spatially is, at its roots, a product of evolution based on survival. Because tactics for survival were a major factor in the evolution of the higher animals, and intelligence in fact developed in the apes and to an even further extent in hominins, it can be said that the uncountable numbers of our ancestors and their brethren that served as prey for our enemies during the long course of evolution leading to the human contributed to our existence today. Furthermore, we can probably think of self-consciousness as a trait that developed because it was advantageous for survival to monitor one's situation continually and in detail.

Thinking = Simulation Looked at in this way, "thinking" is a kind of simulation done in the head, and can be said to be similar to simulations that are being actively performed on computers these days. "Truths" and "natural laws" obtained by research in the natural sciences correspond, in this author's opinion, to subroutines that generate correct results in simulations using computers.

Therefore, it is safer not to be overly obsessed by such concepts as "absolute truth". (The late) Dr. Asajiro Oka, a pioneer of evolutionary studies in our country, compared the cerebra of various vertebrates from fish to the human in his book *Evolution and Life* (1906) and writes "philosophers and their like acknowledge that only their brains are absolutely perfect, and try very hard to speculatively discern the truth of universe; but if we consider all mankind in light of the course of cerebral evolution, there is not much difference between an ignorant, superstitious person and a famous philosopher; and although some difference exists between the two, they are both the same in that they can never be perfect because they have arisen from the same ancestor, have progressed in the same direction, and are in the process of progressing further". These are apt words to caution against speculative runaway.

Development of Social Life In addition to the development of intelligence, collectivization, that is the development of social life, played an important role in the process of human evolution. Humans, who are helpless against fierce beasts when alone, were able to confront strong adversaries because they gathered together and formed societies; and by collaborating in large numbers were able to hunt large prey.

On the other hand, high population densities created difficult problems. One was infectious diseases. In response to this, genetically determined resistance appears to have developed by selection. Furthermore, group living produced leaders, which increased the effectiveness of the group, and the ability of a member to become a leader when necessary must naturally have developed. In particular, the human psychological disposition to trust in one great authority and to submit to it was likely very useful in unifying the tribe. In this way, through the process of intergroup selection, these properties were established as genetic traits and became increasingly powerful.

In my opinion, the disposition to believe in religion is also a product of group living. So-called "patriotism" (it may be more appropriate to call it loyalty to one's group) may have developed through intergroup selection, because such a psychological tendency was advantageous in selection between tribes. In any case, the genetic constitution that forms the foundation for our physical and mental properties are the products of such evolution.

Difficulties of Civilized Society Subsequently, with the development of civilization, the rate of progress of civilization came to supersede the rate of biological progress. This tendency became even more pronounced with the development of natural science. It is well known that the resultant modern civilized society is plagued by many difficult problems. Among these, nuclear weapons and bacterial weapons, born of scientific and technological progress, are major threats to human survival itself. Also, although not as threatening, explosive population growth is a problem requiring prompt attention.

On the assumption that these urgent and serious problems are solved in the near future and that the human is assured a distant future, I would like to take a little time to think about the question of eugenics, including topics on the biological improvement of the human species. In fact, it has often been pointed out from the standpoint of eugenics that modern civilized society has various problems.

9.2 Thinking about the Question of Eugenics

Taboo Eugenics was born of the idealism of Francis Galton, Darwin's cousin and one of the founders of biometrics. However, subsequently, it was misused in the nonscientific and non-humanitarian racial policies of Nazi Germany and completely lost its credibility. As a result, the question of eugenics is now almost taboo, and a tendency has arisen to regard even its discussion as evil. Nevertheless, from the standpoint of biological evolution, this is a question that cannot be avoided in thinking about the distant future (in units of tens of millennia) of mankind.

Medical Progress and Reduction of Selection In thinking about the question of eugenics, what immediately comes to mind is the damage due to the accumulation of mutations. With the drastic decrease in mortality and the curing of infertility due to medical progress, and moreover when family planning is fully achieved, the difference in the number of children contributed to the next generation by different couples will be reduced, and natural selection in the strict sense will gradually decrease. This entails that the removal of mutations will gradually become difficult. For example, when we look at infant mortality in the first year after birth in Japan, this used be 150 per 1000 in 1900, but it had decreased to 15, i.e., one tenth the previous value, by 1967. In addition to the change in mortality, quantitative studies have shown that for fertility also, the possibility for natural selection to act is decreasing. According to a study in the United States that uses Crow's index of total selection, which was mentioned in the latter half of Chap. 5, the average number of children decreased temporarily from 1910 and increased a little from about 1935, but the index that shows the possibility for natural selection to act has ultimately decreased. In short, when the number of children produced by couples becomes similar, the possibility for selection to act is reduced, and the index of total selection expresses this quantitatively.

Gene mutations at the molecular level include substitutions, deletions, transpositions, duplications, etc. of DNA bases; it is significant that, excluding the

Table 9.1 Mutation rates in the human

Mode of inheritance	Disease	Rate (per generation)
Autosomal dominant	Achondroplasia (limb shortening)	5×10^{-5}
Sex-linked	Hemophilia	3×10^{-5}
Autosomal recessive	Albinism	3×10^{-5}
	Microcephaly	4×10^{-5}
	Phenylketonuria	3×10^{-5}
Chromosomal abnormality	Down syndrome (Mongolism)	1/700
	Total including others	About 1% of births

neutral cases, they tend to undermine rather than improve the function of genes. The fundamental reason for this is, that as noted in Chap. 4, mutations are random events at the molecular level and reduce the information content of the genetic commands written with the four kinds of DNA bases as letters. The damage due to gene mutations will manifest themselves phenotypically in every possible way, but among them what should cause the most concern is the lowering of intelligence. Chromosomal abnormalities are a kind of mutation that also have undesirable phenotypic effects, including the lowering of intelligence. For example, it has already been mentioned that Down's syndrome is caused when an extra copy of a small chromosome enters the fertilized egg nucleus as a result of chromosomal nondisjunction.

Table 9.1 lists a few examples of mutation rates in the human; the gene mutation rates for hemophilia and albinism are approximately 3 in 100,000 per gene locus. At first sight this seems a low rate, but the human cell nucleus contains many genes, so that to convert to per gamete we need to multiply this several thousand or several tens of thousand times. In addition, there are chromosomal abnormalities, which together are estimated to occur in about 1% of births.

These gene mutations and chromosomal abnormalities (especially those with clear phenotypic effects) are generally disadvantageous for the survival of an individual, and so have, during the long process of human evolution in the past, been maintained at low frequencies in the population in the balance between their removal by natural selection and their production by mutation. However, medical progress entails that the abnormalities caused by these mutations will phenotypically be gradually cured. For instance, phenylketonuria can be detected at birth, it is possible for intelligence to develop normally by providing a diet that restricts phenylalanine, and this treatment is in actual use. This is to be welcomed from a humanitarian standpoint, but harbors a big problem eugenically in that the mutant gene will be transmitted to the next generation (in the long run, advances in treatment methods will entail that the number of people that must avail themselves of these advances in treatment methods will increase). However, the rate of increase in the frequencies of individual genes will be very small, and we need not worry too much.

Degeneration of the Genetic Commands What is more serious is the problem of mutations as a whole. This is the problem where the majority of genes that in the past were deleterious are rendered selectively neutral by medical progress, undergo neutral evolution under mutation pressure, and come to be fixed in the population. Dealing with the damage due to the accumulation of such mutations only by such superficial measures as simply ameliorating the environment, or repairing the phenotype (the majority of medical treatments belongs to this category), or developmental control may be possible for a while, but is not feasible in the long run and will result in much waste socially.

Here "developmental control" refers to what the geneticist J. Lederberg once named euphenics. This is the possibility of artificially controlling development such that a person with a genetic defect can be equal to a normal person. It may of course not be possible to do this now, but it was proposed in the belief that in the future this might work better than manipulating genes.

However, at the root of all human existence is the human as a biological organism; and when we consider that its form is determined by translation of the genetic commands residing in the nucleus of the fertilized egg, to permit the degeneration of these commands is ultimately to bring about the degeneration of mankind. Although it may not become a problem for tens of millennia, in order that humans be able to devote most of their intellectual, labor, and material resources to constructive and promising projects rather than various phenotypic measures, eugenic steps would seem to be absolutely necessary.

Negative Eugenics Eugenics has both negative and positive aspects. Let me first describe negative eugenics which has as its goal limiting the accumulation of deleterious mutations below a certain level.

Various methods can be thought of and have been argued, but the makeshift measure of avoiding marriages between carriers is fundamentally useless. Nevertheless, it has already become possible to detect carriers of various genetic diseases in the heterozygous state, making it possible to avoid marriages between persons who are heterozygous for the same recessive factor, and this method will be effective for a while. However, as the mutant gene eventually increases, suitable marital combinations will themselves become impossible.

Ultimately, the two measures that are realistically effective would seem to be for people with more than the average number of deleterious genes to in some way limit the number of their children, or for fertilized eggs that are known to carry deleterious mutant genes to be removed early in development. In particular, it would seem desirable for both the individuals involved and for society as a whole not to permit fertilized eggs containing chromosomal abnormalities to develop. Very recently, by using the so-called method of amniocentesis, prenatal detection of chromosomal abnormalities and their removal by abortion has become possible, and this is good news. The American geneticist Bentley Glass once said that a time will come when to be born healthy will be regarded as one fundamental human right comparable to the right of each individual to receive an education.

There is something I would like to add in relation to this. There are geneticists who even now, in emphasizing that the existence of genetic variation is in itself desirable for mankind, forget that not all variation falls in the normal range, but also includes recessive lethals and causes of terrible diseases and is not uniformly desirable. This author cannot agree with the notion that large variation in itself is a good thing.

Dangers and Limitations of Gene Manipulation Furthermore, as to the possibility of completely replacing a defective gene by a normal gene or of performing molecular microsurgery, we cannot, due to the remarkable progress in present day gene manipulation techniques, ignore the possibility that in the comparatively near future such methods will be realized with respect to some genetic diseases. However, the dangers that must be considered when this possibility is realized will be much more serious than the present-day problem of the safety of medicines. Moreover, with the exception of special cases, there will be major limitations in terms of cost as a general method of treatment. In particular, if this method is used only on genes of somatic cells and is not applied to germ cells, the effect lasts only for one generation and it becomes necessary, as with phenotypic measures, to repeat it every generation.

9.3 Positive Eugenics and the Future of Mankind

Subtle Interactions Between Many Genes Next, I would like to take up the question of positive eugenics, which aims to amplify characters that are thought to be desirable for mankind. For this purpose, microscopic surgery of genes mentioned in the previous section is a not very useful method. The principle reason for this is that the traits that are the focus of positive eugenics, such as general intelligence, health, social cooperativeness, etc. depend on the subtle interactions of many genes, individual differences are large, and cannot be improved by replacing a specific defective gene by the normal type.

Even if in the distant future it will become possible to freely replace DNA base pairs in the nucleus, it is premature to think that the genetic qualities of the human can be freely improved. This is comparable to the notion that the blueprint for any splendid machine can be drawn provided paper and a pencil are available, and ignores the higher dimensional problem of information. Moreover, the conundrum exists of how far the human can improve its brain by using its brain.

An Issue Relevant to the Essence of Being Human Recently, on the foundations of molecular biology and other modern sciences, so-called biotechnology has appeared with a flourish, and this will bring about unimaginably large changes in our lives from now on. With such a trend in mind, and from a long term perspective, I would like to describe two ideas that have been proposed to date as measures for positive eugenics.

What I would like to make clear at the beginning is that the measures of positive eugenics that I will describe below received severe opposition and criticism from many people simultaneously with their proposal, and that it is definitely not the case that I am in total agreement. Because it is an issue that bears on the essence of being human and on the core of humanity, it is natural that a great controversy has ensued. Not a few critics did not even try to understand the true intentions of the advocates and adopted a tenor of total rejection.

Nevertheless, when we try to think about the long term future of mankind spanning many millennia, many tens of millennia, or even longer, I believe these eugenic measures are sufficiently worthy of consideration. Thus, with the intention of offering basic material for the discussion of this question, and for the moment setting aside social judgment and appraisal, I would like to describe proposals that I believe are important from an evolutionary genetics standpoint.

Muller's Plan for a Sperm Bank The first idea is the method of germ plasm selection, utilizing a sperm bank, proposed by the American geneticist H. J. Muller. Muller is an eminent scholar who is judged by some to be the greatest geneticist of this century; in his later years, he put his heart into genetically improving mankind and wrote many papers on this subject. As a result, he was exposed to severe opposition and criticism from many people.

Muller's plan was to establish a sperm bank; to freeze and store the sperm of many people there; and several decades after their deaths, when judgments of their personalities, health, and especially their achievements could be made objectively, to choose sperm for use in artificial insemination. The final decision would be made by the recipients, but Muller suggested that to begin with, this be used by couples that could not have children because the husbands were infertile. Here, what is quite at variance with the prevailing social mood is that the name of the sperm donor is not concealed, but that the parents of the child who is born would openly and with pride reveal this name. According to Muller's thinking, if it becomes clear that a child born in this way is on average superior, this method will spread, and genetic effects would be achieved.

This plan of Muller's of course met with severe opposition from various quarters. However, considering that social morals are not absolute, it is not unthinkable that in the distant future there will be a society that will accept this. In particular, when in the future humans set about colonizing space and advance beyond the solar system into the Milky Way, Muller's plan may unexpectedly become a reality.

If we think about various aspects comprehensively, Muller's method of germ plasm selection may, as a positive eugenic measure for humans, perhaps be the safest, most reliable, and in the long term most effective method, scientifically speaking, for genetically improving such traits as general intelligence, health, and social cooperativeness.

Nevertheless, with regard to this method, there is one caution that I must add from the standpoint of population genetics. This is that the population of sperm donors should not be too small. If this is too small, not only will the possibility of inbreeding immediately arise, but the genetic variability of the population will be limited, and there is a danger that what is called the selection limit in breeding will be lowered. Incidentally, the theory of selection limits was developed by A. Robertson in

England, based on my theory of the fixation probability of genes in a finite population, and is a theory that deals with how selection should be applied for long term breeding to be effective. From this standpoint also, it is extremely bad to choose only a handful of extremely gifted geniuses (for example Nobel Prize winners) as parents. Therefore, when we think about an appropriate method from the standpoint of population genetics, it is desirable to let the effective number of the population of actual sperm donors per generation be suitably large (for example one million persons), and that the proportion of the population of sperm donors relative to the whole be not too small. For example, if we let this be 1%, then if the total number of individuals in the population is 100 million, it is better to take about one million persons who by some criteria are considered superior to produce the next generation. I should add that this calculation is of course only a theoretical exercise, of relevance only if Muller's method is socially approved.

Haldane's Cloned Human The second positive eugenic measure that I will take up is the idea of making a clone of a person expressed by J. B. S. Haldane a long time ago. This idea calls for growing the somatic cells of a person with superior qualities and producing many genetically identical individuals.

It is not possible with this method to achieve a long term improvement in the genetic qualities of a population as with germ plasm selection, but if the techniques for growing cells into individuals are perfected, the result is assured. Making a cloned human, together with microsurgery of genes, may perhaps in the future become a measure to supplement germ plasm selection. However, I would like to emphasize that many scholars have already expressed the opinion that making a cloned human is unacceptable, because it is blasphemy against humanity or violates the dignity of the individual.

In my opinion, in the distant future, when extrasomatic cultivation of a fertilized egg comes to widely replace natural birth for the purpose of liberating women from the inconvenience of pregnancy and childbirth, the psychological effects of germ plasm selection will diminish. Extrasomatic cultivation of the egg may, in addition to eugenic treatments, become useful for controlling immunological responses in preparation for organ transplants after development.

9.4 Human Expansion into Space and Evolution

End of the Beginning The fate of mankind on this Earth is in the hands of humans themselves. Moreover, as far as the Earth is concerned, the time when human evolution can be left to follow its natural course is likely already over. If mankind overcomes the current crisis of survival, they will eventually set about colonizing space from the Earth as their base. Winston Churchill once famously said "the end of the beginning", and our times may be like that. In fact, that is how I would like it to be.

Space Colony Already more than 10 years ago, Gerald K. O'Neill of Princeton University proposed, for the purpose of the colonization of space, constructing a large cylindrical space ship with a diameter of four miles (approximately 6.6 km) and a length of 16 miles (approximately 26 kilometers). The interior is filled with air, and because the space ship rotates around its longitudinal axis once approximately every 2 min, centrifugal force is generated, and people are able to walk upright as on the Earth. In the interior, there are three valley-like parts and three windows, and it is possible with giant wing-like mirrors to produce daily and seasonal changes. The plan is to launch it to the Lagrange point between the Moon and the Earth where gravity is cancelled. The first such space colony would be assembled mainly with materials from the Lunar surface and about 2% with materials transported from the Earth. Life on this spaceship is self-sufficient, and in the valley-like places where people live would be arranged pastures, fields, mountains, forests, lakes, and villages. The time may come when innumerable such space colonies float like islands within the solar system.

Several million people can live comfortably on this space colonization island. What is important is that this space colony can be constructed from steel and aluminum using current technology. The cost is of course tremendous, but in recompense, it is possible to build a satellite solar generating plant here, to transmit the electricity to the Earth after converting it to ultrashort wave, and to relieve the power crisis there.

Arthur Clarke, my favorite science fiction author, has written a novel *Rendezvous with Rama* (Japanese translation *Space Rendezvous*, Hayakawa Shobo, 1979). This is a masterpiece among recent science fiction, and is based on ideas similar to O'Neil's space colony. The spacecraft called Rama that appears in this novel is a cylinder with a diameter of 20 kilometers and a length of 50 kilometers, and cities and rivers are arranged in its interior. This is a story in which the spacecraft travels through space for many millions of years, replenishes materials and energy from the solar corona as it passes through the solar system, and eventually flies off in the direction of the distant Magellanic Clouds.

Does Extraterrestrial Intelligence Exist? It has been claimed that uncountably many planets with advanced civilizations exist in our universe. And that if there is a communication network between them, it should be possible by checking radio waves to detect the existence of civilizations in the universe; the search for extraterrestrial intelligence (abbreviated SETI) is a project based on this idea.

However, I am highly doubtful of the claim that there are many worlds (other than the Earth) with advanced civilizations in the Milky Way. In the calculations of the astronomers that lead to such claims, the estimate of about 10^{11} stars in the Milky Way is probably secure, and it may be correct that the probability is fairly large (for example, a few percent or greater) that planets exist with an environment in which biological organisms can exist. However, the notion that given a "suitable" physical environment, the probability that life will emerge there in 3–4 billion years (let this be P_a), and the probability that life, once arisen, will evolve toward highly intelligent organisms like the human (P_b) are both fairly large seems to be groundless. Writings by scholars in fields related astronomy often and facilely set the values of P_a and P_b close to one.

In retrospect, in the Middle Ages before the development of science, the Earth was thought to be the center of the universe. Later, by the Copernican revolution, it came to be recognized that the Earth and the solar system to which it belongs are average and peripheral entities in the universe. However, it is a leap in logic to think that the existence of an intellectual life form like the human is also an average occurrence in the universe. Clearly, physical laws hold in the same way on the Earth and on a distant star many hundreds of millions of light years away. Hence, there are sufficient grounds for thinking that the Earth and solar system are average physical entities. On the other hand, we can say that there is no law of life that holds throughout the universe. As I have explained repeatedly in this book, it is an important point how greatly biological evolution has been swayed by chance. Moreover, from the standpoint of the neutral theory, each and every allele that we carry is the product of unbelievably good fortune. This author feels that if the human is a miracle in the universe which arose by the concatenation of extreme good fortune, it is not surprising if the value of P_aP_b is 10^{-100}. Therefore, the claim that there are no advanced intellectual life forms in the Milky Way other than the human would seem to be far more reasonable.

In this connection, I would like to introduce the following interesting calculation (this was prompted by a question once raised by the physicist E. Fermi, to the effect that, if spacemen exist, where would they be?). If starting from one planet in the Milky Way with an advanced civilization, migrating to the next star 10 light years away in a spaceship that can fly at one tenth the speed of light ($0.1c$), migrating to the next star from this as its base, and with an interval of 500 years between one migration and the next migration, it would take only 5 million years to spread throughout the Milky Way. Since there is no evidence that spacemen have visited the Earth, this may support the argument above that there are no intelligent organisms other than the human in the Milky Way.

Altruism In thinking about the human colonization of space, a valuable human asset that should be mentioned in addition to an advanced intelligence is "altruism" (loving others is the most splendid of the human dispositions). It can perhaps be said that the altruism of the human may be healthy and show hope for the future, in that it is not blind as in the bee, but is a complex affair that includes individual judgment and is often done in expectation of future reciprocation with consideration of profit and loss.

Superhuman In many futuristic theses written by zoologists, humans will have gone extinct in the not distant future, after which rodents will infest the Earth in place of the human. I would, however, like to conclude this book, not with such a pessimistic view of the future, but rather dreaming of humans employing science and technology and cooperating to expand into the universe.

F. J. Dyson once argued the possibility of enveloping a star with a spherical surface composed of matter obtained by breaking up a planet, and creating a civilization inside (this is well known as the "Dyson sphere"); it may be many million years before such worlds are created by the descendants of mankind everywhere in the Milky Way. When this is realized, adjacent colonies will be separated by many light years of space, so that there will be a strong isolating action, and it is unavoidable that genetic differentiation will occur. And in one world, positive eugenics may be put into practice according to Muller's plan described in the previous section, and superhumans may be born. Our descendants born in this way with an ultrahigh intelligence may solve mysteries of the universe that are far beyond our current understanding. Nevertheless, even in this world of superhumans, many of the superior physical (genetic) natures and cultural inheritances possessed by today's humans will be preserved. This is just like the letters of the alphabet that began with one tribe living in the Sinai Peninsula three thousand and several hundred years ago, which are now widely used throughout the world as an indispensable tool for embodying information.

Privilege of Being Human Returning to reality we are all, to some extent, cog wheels incorporated into a large social structure, and busily lead our daily lives. An academic discipline like evolutionary theory which is not of much practical use in such lives, let alone in making money, tends to be unappreciated. However, being aware of the descent of mankind and imagining their future is a privilege permitted only to humans. And evolutionary research is an important academic field in that it aids in such contemplation. It can be said in this regard what a wonderful thing it is that, although each of us is temporally and spatially an infinitely small entity compared to the universe as a whole, we can hold the entire universe inside our heads through the processes of thought.

References

Chapter I

Dobzhansky T (1970) Genetics and the evolutionary process. Columbia University Press, New York

May RM (1986) How many species are there? Nature 324:514

Saunders SR (1985) The inheritance of acquired characteristics: a concept that will not die. In: Godfrey LR (ed) What Darwin began. Allyn and Bacon, Boston, p 148

Simpson GG (1964) This view of life. Harcourt, Brace & World, New York

Chapter II

Haldane JBS (1932) The causes of evolution. Harper & Row, London

Provine WB (1971) The origins of theoretical population genetics. University of Chicago Press, Chicago

Provine WB (1986) Sewall Wright and evolutionary biology. University of Chicago Press, Chicago

Sturtevant AH (1965) A history of genetics. Harper & Row, London

Chapter III

Alvarez LW (1983) Experimental evidence that an asteroid impact led to the extinction of many species 65 million years ago. PNAS 80:627

Calder N (1983) Timescale. Viking Press, New York

Crompton AW, Taylor CR, Jagger JA (1978) Evolution of homeothermy in mammals. Nature 272:333

Lieberman P (1984) The biology and evolution of language. Harvard University Press, Cambridge

Raup DM, Sepkoski JJ Jr (1986) Periodic extinction of families and genera. *Science* 231:833

Romer AS (1968) The procession of life. Weidenfeld and Nicolson, London

Schopf JW (ed) (1983) Earth's earliest biosphere: its origin and evolution. Princeton University Press, Princeton

Valentine JW (1985) The evolution of complex animals. In: Godfrey LR (ed) What Darwin began. Allyn and Bacon, Boston, p 258

M. Kimura, *My Thoughts on Biological Evolution*, Evolutionary Studies,
https://doi.org/10.1007/978-981-15-6165-8

Chapter IV

Furusho T (1974) Genetic study on stature. Jap J Human Genet 19:1
Goodenough U (1984) Genetics, 3rd edn. Saunders College Publishing, Philadelphia
Seno T (1962) Genetical studies on copper resistance of *Saccharomyces cerevisiae*. Jap J Genet 37:207

Chapter V

Barlow N (1958) The autobiography of Charles Darwin. Collins, London
Darwin C (1859) The origin of species by means of natural selection. John Murray, London
Dawkins R (1986) The blind watchmaker. Longman Scientific & Technical, London
Endler JA (1986) Natural selection in the wild. Princeton University Press, Princeton
Fisher RA (1930) The genetical theory of natural selection. Oxford University Press, Oxford

Chapter VI

Crow JF, Kimura M (1970a) An introduction to population genetics theory. Harper & Row, London
Kimura M, Ohta T (1971) Theoretical aspects of population genetics. Princeton University Press, Princeton

Chapter VII

Crow JF, Kimura M (1970b) An introduction to population genetics theory. Harper & Row, London

Chapter VIII

Dyson FJ (1985) Origin of life. Cambridge University Press, Cambridge
Jukes TH (1985) A change in the genetic code in *Mycoplasma capricolum*. J Mol Evol 22:361
Takahata N (1986) Coalescent processes and diffusion processes in population genetics. Math Sci 272:34
Yamao F, Muto A, Kawauchi Y, Iwami M, Iwagami S, Azumi Y, Osawa S (1985) UGA is read as tryptophan in *Mycoplasma capricolum*. PNAS 82:2306

Chapter IX

Bracewell RN (1974) The galactic club. W.H. Freeman, San Francisco
Finney BR, Jones EM (1985) Interstellar migration and the human experience. University of California Press, California
Heppenheimer TA (1977) Colonies in space. Warner Books, New York

The manufacturer's authorised representative in the EU is Springer Nature Customer Service Centre GmbH, Europaplatz 3, 69115 Heidelberg, Germany. If you have any concerns regarding our products, please contact ProductSafety@springernature.com

Printed and bound by CPI Group (UK) Ltd, Croydon, CR0 4YY
30/09/2026
02225266-0001